DYNAMICS AND CHAOS IN MANUFACTURING PROCESSES

WILEY SERIES IN NONLINEAR SCIENCE

Series Editors: **ALI H. NAYFEH, Virginia Tech**
ARUN V. HOLDEN, University of Leeds

Abdullaev	Theory of Solitons in Inhomogeneous Media
Bolotin	Stability Problems in Fracture Mechanics
Kahn and Zarmi	Nonlinear Dynamics: Exploration through Normal Forms
Moon (ed.)	Dynamics and Chaos in Manufacturing Processes
Nayfeh	Method of Normal Forms
Nayfeh and Balachandran	Applied Nonlinear Dynamics
Nayfeh and Pai	Linear and Nonlinear Structural Mechanics
Ott, Sauer, and Yorke	Coping with Chaos
Pfeiffer and Glocker	Multibody Dynamics with Unilateral Contacts
Qu	Robust Control of Nonlinear Uncertain Systems
Vakakis et al.	Normal Modes and Localization in Nonlinear Systems
Yamamoto and Ishida	Linear and Nonlinear Rotor Dynamics: A Modern Treatment with Applications

DYNAMICS AND CHAOS IN MANUFACTURING PROCESSES

Edited by

FRANCIS C. MOON

A Wiley-Interscience Publication

JOHN WILEY & SONS, INC.

New York • Chichester • Weinheim • Brisbane • Singapore • Toronto

This text is printed on acid-free paper.

Copyright © 1998 by John Wiley & Sons, Inc.

All rights reserved. Published simultaneously in Canada.

Library of Congress Cataloging in Publication Data:

Dynamics and chaos in manufacturing processes / edited by Francis C.
   Moon.
        p.   cm. -- (Wiley series in nonlinear science)
     "A Wiley-Interscience publication."
     Includes bibliographical references and index.
     ISBN 0-471-15293-5 (alk. paper)
     1. Manufacturing processes.   2. Chaotic behavior in systems.
   I. Moon, F. C., 1939–   II. Series.
   TS183.D885   1998
     670--dc21   96-29975

Printed in the United States of America

10 9 8 7 6 5 4 3 2 1

CONTENTS

CONTRIBUTORS

Ernest Barreto
Institute for Plasma Research
Department of Physics
University of Maryland
College Park, MD 20742

B. S. Berger
Department of Mechanical
Engineering
University of Maryland
College Park, MD 20742

Harish P. Cherukuri
Department of Mechanical
Engineering and
Engineering Sciences
The University of North Carolina at
Charlotte
Charlotte, NC 28223

Char-Ming Chin
Department of Engineering Science
and Mechanics
Virginia Polytechnic Institute and
State University
Blacksburg, VA 24061-0219

David J. Christini
NeuroMuscular Research Center and
Department of Biomedical
Engineering
Boston University
Boston, MA 02215

James J. Collins
NeuroMuscular Research Center and
Department of Biomedical
Engineering
Boston University
Boston, MA 02215

Matthew A. Davies
Automated Production Technology
Division
Manufacturing Engineering
Laboratory
National Institute of Standards and
Technology
Gaithersburg, MD 20899-0001

B. F. Feeny
Department of Mechanical
Engineering
Michigan State University
East Lansing, MI 48824-1226

Celso Grebogi
Institute for Plasma Research
Department of Mathematics
Institute for Physical Science and
Technology
University of Maryland
College Park, MD 20742

Mark A. Johnson
General Electric Corporate Research
and Development
Schenectady, NY 12306

Robert E. Johnson
Department of Mechanical
Engineering and
Engineering Sciences
The University of North Carolina at
Charlotte
Charlotte, NC 28223

Ying-Cheng Lai
Department of Physics and
Astronomy
Department of Mathematics
Kansas Institute for Theoretical and
Computational Science
The University of Kansas
Lawrence, KS 66045

Dung T. Le
Department of Mechanical
Engineering and
Institute for Systems Research
University of Maryland
College Park, MD 20742

Paul S. Linsay
Plasma Fusion Center
Massachusetts Institute of
Technology
Cambridge, MA 02139

Michael Marder
Department of Physics and
Center for Nonlinear Dynamics
The University of Texas at Austin
Austin, TX 78712

Ioannis Minis
Department of Mechanical
Engineering and
Institute for Systems Research
University of Maryland
College Park, MD 20742

Francis C. Moon
Department of Mechanical and
Aerospace Engineering
Cornell University
Ithaca, NY 14853

Ali H. Nayfeh
Department of Engineering Science
and Mechanics
Virginia Polytechnic Institute and
State University
Blacksburg, VA 24061-0219

Stanley J. Ng
Department of Mechanical
Engineering and
Institute for Systems Research
University of Maryland
College Park, MD 20742

Jon Pratt
Department of Engineering Science
and Mechanics
Virginia Polytechnic Institute and
State University
Blacksburg, VA 24061-0219

Steve W. Shaw
Department of Mechanical
Engineering
Michigan State University
East Lansing, MI 48824-1226

Gábor Stépán
Department of Applied Mechanics
Technical University of Budapest
Budapest
H-1521 Hungary

A. Galip Ulsoy
Department of Mechanical
Engineering and
Applied Mechanics
University of Michigan
Ann Arbor, MI 48109-2125

Guangming Zhang
Department of Mechanical
Engineering and
Institute for Systems Research
University of Maryland
College Park, MD 20742

PREFACE

This book explores the dynamics, chaos, and complexity in manufacturing processes. Nonlinear dynamics phenomena in material processing have long been recognized in turning, milling, grinding, rolling, and other machining and forming processes. But until recently the mathematical and experimental tools to diagnose and understand the dynamics have been lacking. This book brings together experts in machining and forming processes as well as in nonlinear and chaotic dynamics. The editor and the authors intend this book to serve as an introduction to the phenomena, to provide a source for the past and emerging literature, and to present some new methods to diagnose and control nonlinear dynamics in manufacturing processing.

The eight chapters in Part I explore the direct application of nonlinear dynamics and chaos theory to machining, grinding, and rolling processes. Many of these chapters discuss the difficult problem of chatter dynamics in cutting processes. These chapters illustrate three generic dynamic subproblems: The stability of cutting processes, subcritical bifurcations, and chaotic vibrations.

The chapters in Part II introduce new ideas in nonlinear dynamics that may have direct application to manufacturing processes. These include impact and friction dynamics, fracture dynamics, and new control methods using the theory of chaotic dynamics.

The material for this book came out of a workshop sponsored by the Institute of Mechanics and Materials (IMM) at the University of California at San Diego (UCSD), LaJolla, March 20–22, 1995. The IMM is supported by the National Science Foundation as a venue to bring together researchers at the boundary between the disciplines of mechanics and materials. The Institute runs over a dozen workshops a year on many interdisciplinary subjects. This book is dedicated to Professor Richard Skalak of UCSD who directed the IMM from 1991 to 1996. His broad interests and commitment to interdisciplinary research have been a guiding light for the Institute. His personal research interest and leadership in bioengineering illustrate his devotion to tackling problems at the boundaries of different sciences. Professor Skalak taught at Columbia University for several decades before coming to UCSD to head the IMM. The editor wishes to express his debt to Professor Skalak for his recognition of the importance of dynamics in material processing and for enthusiastically encouraging and supporting the workshop and this book.

Thanks are also due to Ms. Donna Shinn of the IMM for her work in organizing the workshop. The editor further wants to thank Professor Henry Abarbanel of the Center for Nonlinear Dynamics at UCSD for his support and collaboration. A special note of thanks is due to Ms. Debbie DeCamillo for her typing and help in dealing with all the authors. Finally the authors wish to thank Mr. Greg Franklin and Mr. John Falcone of John Wiley & Sons for their patience in shepherding this book through the publication process.

NONLINEAR DYNAMICS IN MANUFACTURING PROCESSES

1

NONLINEAR DYNAMICS AND CHAOS IN MANUFACTURING PROCESSES

F. C. MOON and M. A. JOHNSON

1.1 INTRODUCTION

This book will explore dynamical phenomena in manufacturing and material fabrication processes. It will also introduce new analysis tools from the field of nonlinear and chaotic dynamics. Emphasis is placed on the application of these analysis tools to manufacturing processes. The interest in dynamics and manufacturing is prompted by the development of high-speed spindles in machine tools with faster cutting speeds and feeds (e.g., see Ashley, 1995). Also the search for new control ideas in material processing requires the understanding of dynamic models of the manufacturing process.

The creation of most manufactured objects involves some dynamic processes. It has long been known that vibrations and noise accompany cutting, grinding, drawing, rolling, and other traditional manufacturing processes (Figure 1.1). As early as 1907 F. Taylor, in his presidential lecture to the ASME, commented on a particular form of vibration in the cutting of metals known as chatter.

> Chatter is the most obscure and delicate of all problems facing the machinist — probably no rules or formulae can be devised which will accurately guide the machinist in taking maximum cuts and speeds possible without producing chatter.

Dynamics and Chaos in Manufacturing Processes, Edited by Francis C. Moon
ISBN 0-471-15293-5 © 1998 John Wiley & Sons, Inc.

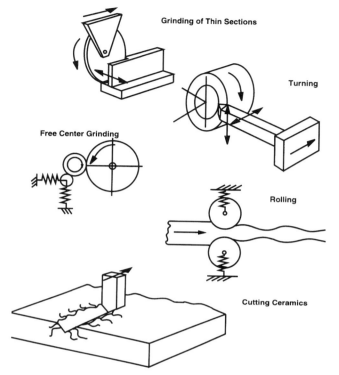

FIGURE 1.1. Manufacturing processes with significant dynamics effects.

The numerous dynamic phenomena in manufacturing processes include:

- chatter in the machining of metals and ceramics,
- dynamically induced thickness variations in the rolling of sheet metals,
- unsteady forces in the extrusion and drawing of metals and polymers,
- web and string dynamics in fiber, textile, and paper manufacturing.

We now know that nonlinear dynamics and even chaos can accompany processes that involve material flow. The fundamental mechanics in material flow processes include:

- friction,
- fracture,
- elastoplastic deformation,
- granular flow,
- intermittent impact between materials.

The chapters in this book illustrate two new trends in dynamics research: (1) the advancement of the science and mathematics of nonlinear dynamical systems and (2) the application of advanced dynamics and control tools to manufacturing processes. By bringing together reviews from these two research areas, we hope to encourage new applications of nonlinear dynamics and control to manufacturing technologies. It is hoped that new models for simulation and understanding dynamic material processes as well as new methods of controlling these processes will emerge from the synthesis of these two trends.

A related emerging area is the use of numerical finite element and boundary element methods to simulate the dynamics of manufacturing processes (e.g., see Marusich and Ortiz, 1995). However, this book will not treat this exciting new area.

1.2 WHAT'S NEW IN NONLINEAR DYNAMICS?

This book's premise is that there are new analytic and experimental tools and control ideas from two decades of research in nonlinear dynamics that have direct application to manufacturing processes. The successful application of these new tools requires both a detailed understanding of specific manufacturing processes and some skill in applying the new dynamics. Manufacturing processes involve three-dimensional distributions of deformation strains and temperature and sometimes solid state chemical reactions. Detailed mathematical models involve nonlinear partial differential equations whose solution involves high-dimension numerical techniques. Progress in nonlinear dynamical systems, or so-called *chaos theory*, has shown that new techniques can be applied to develop low-dimensional models that can capture essential features of the dynamics of complex physical systems. These low-dimensional models also provide a conceptual platform to apply control to these systems. The new ideas that have emerged in dynamics include (see also Moon, 1992) the following:

- Patterns of the evolution of behavior from simple to complex, such as from periodic to chaotic motions.
- Sensitive dependence on initial conditions.
- Fractal properties of spatial distributions of orbits in phase space and fractal surfaces in the objects that result from these dynamics.
- The possibility of both spatial and temporal complexity of motion.

The new dynamics has spurred new techniques of analysis, signal processing, and experimental methods, including:

- Poincaré maps of orbits in a phase space.

- Analysis of global bifurcations of the motion.
- Phase space reconstruction from time series data.
- Methods to determine the dimension of the phase space, and attractors in this phase space, using Lyapunov experiments, fractal dimension, and false nearest neighbors.
- Methods to control chaos or to excite chaos when desirable.

Some chapters will provide an introduction to these new methods of analysis, while others will illustrate both potential and successful applications to manufacturing processes.

1.3   DYNAMICS AND SURFACE QUALITY

The creation of any surface is a dynamic process, and the nature of the dynamics should affect the geometric and dimensional quality of the surface. An example is given in Figures 1.2 and 1.3 where an aluminum disc was cut on its face. Both the surface quality and dynamics of the cutting tool were measured (Johnson, 1996). The smoother surface was generated by randomlike dynamics of the cutting tool (we believe it was chaotic), whereas the rough surface was generated while the tool motion was nearly periodic (traditionally known as chatter). It is conceivable that future specification of surface quality will require proscribed cutting dynamics through either active control or passive time history shaping. The intentional dynamic generation of desired surface qualities is not possible today because we do not have accurate and efficient models for many of the most basic manufacturing processes. While numerical finite element and boundary element codes have made important strides in recent years, their application to the design of manufacturing control systems remains a challenge. Low-order models that capture the principal features of the process appear to be of immediate usefulness in the control and design of processes.

1.4   MACHINING DYNAMICS OF METALS AS A PARADIGM

In other chapters, the dynamics of grinding, milling, and rolling processes will be discussed in detail. In the remainder of this chapter we review the problem of chatter and pre-chatter vibrations in turning material-cutting processes. We see this problem as a paradigm for looking at nonlinear dynamics in other manufacturing processes. It illustrates both the promise and problems that await the researcher in applying nonlinear dynamics to material processing.

Tobias' 1965 book is cited by Welbourn and Smith (1970) as "... the pioneering book in the field of chatter." It contains a comprehensive summary of the theoretical and experimental work done on machine tool chatter, including drilling, milling, grinding, and turning, prior to 1965. Tobias echoes Taylor's sentiments on the elusive nature of chatter:

> The physical causes underlying the mechanism are still not fully understood, and this is why it is so often extremely difficult, despite a diagnosis which accords with the facts, to find any remedy short of reducing metal removal rates with consequent lowering of output. In addition, chatter is so inconsistent in character that the tendency of a machine to exhibit chatter effects is often not observed during the development stage.

Early studies of chatter include Arnold (1946) and Doi and Kato (1956). Doi and Kato presented one of the first nonlinear models for chatter dynamics. The survey paper by Tlusty (1978) provides a more recent list of references.

The significance of chatter today is perhaps reflected in the need for increased productivity and efforts at higher metal removal rates. High-speed milling has received considerable attention, yet as noted by Smith and Tlusty (1990) vibration and chatter still impede the realization of the full potential of high-speed milling.

1.5 CHATTER CONTROL

Manufacturers attempt to design machine tools with sufficient stiffness and damping so that tool chatter does not occur. When chatter does occur, system parameters such as the cutting speed, feed rate, depth of cut, tool geometry, and the tool material can be adjusted to avoid chatter. Even ad hoc techniques such as surrounding a tool with clay or grinding a flat on the clearance face of the tool have been used.

Though these ad hoc techniques are often the first line of defense against chatter, there have also been efforts to control chatter through active and passive means. A number of such efforts are presented here to illustrate the diversity of methods used to control chatter, as well as the lack of a generally accepted control strategy.

One of the earliest attempts to control chatter was that of Hahn (1951). Hahn added a small mass to a cavity in the tip of a boring bar. The air film between the mass and the sides of the cavity introduced viscous friction when the tip of the boring bar oscillated. Other efforts have attempted to use electrohydraulic or piezoelectric actuators in feedback control systems. During the 1960s the United States Air Force supported several research efforts at understanding and controlling chatter. Among the results of these efforts are the works of Merrit (1965) and Sisson and Kegg (1969). A summary of a portion of the research carried out at the Cincinnati Milling Machine Company, today known as Cincinnati Milacron, is given by Lemon and Long (1965). An electrohydraulic system with feedback control was given the name of Controlled Mechanical Impedance and is outlined by Comstock, Tse, and

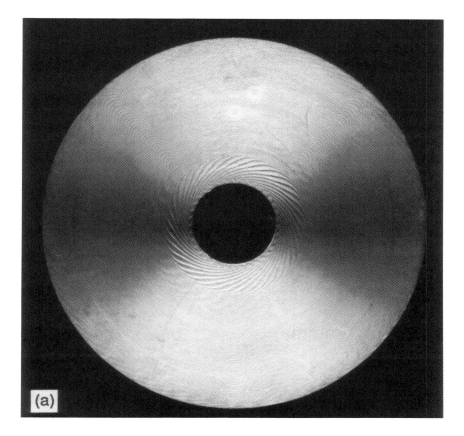

FIGURE 1.2. Face cutting of 6061-TG aluminum disc at 1000 feet per minute cutting speed. (*a*) Chip width 0.01 inches, smooth surface finish accompanied by chaotic dynamics. (*b*) Chip width 0.03 inches, rough surface finish accompanied by periodic dynamics.

Lemon (1968). Another similar attempt at active control was that of Nachtigal and Cook (1970), Nachtigal (1972), and Klein and Nachtigal (1975).

A unique passive control approach was developed by Optiz, Dregger, and Roese (1966). They used irregularly spaced teeth in a milling cutter to reportedly combat the effect of regenerative chatter. A variation of this idea was implemented by Lin, DeVor, and Kapoor (1990).

Delio, Tlusty, and Smith (1992) describe how the sound generated by chatter can be used for recognition of chatter and subsequent control. This idea was implemented in a system developed at Manufacturing Laboratories Incorporated and the University of Florida. The system, labeled the Chatter Recognition and Control (CRAC) system, uses a microphone to monitor a milling process. The strong high-frequency sound typically associated with

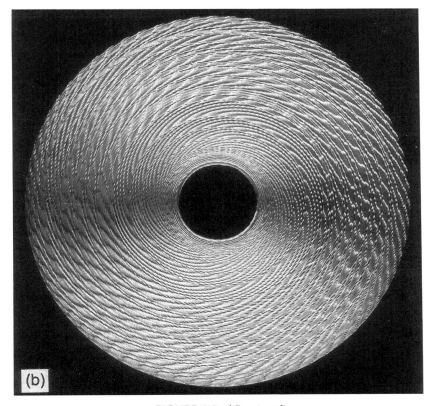

FIGURE 1.2. (*Continued*)

chatter can be recognized by the system. The system halts the milling process once chatter is detected and calculates new speeds and feeds that supposedly avoid chatter.

Chatter control schemes usually rely on some underlying assumptions about a mathematical model for the system to be controlled. In the context of control theory, this is referred to as a model for the plant. The lack of fundamental understanding of machine tool vibrations has led to a large number of models for the plant and no widespread acceptance of any one model. As a result there are a number of control and actuation strategies, with no particular strategy gaining industrial acceptance and widespread implementation.

1.6   MODELS FOR MACHINING DYNAMICS

Differential difference equations, or differential equations with delay, have been widely used as models for regenerative chatter. The principle of regenerative chatter in turning is illustrated in Figure 1.4. A cylindrical workpiece rotates

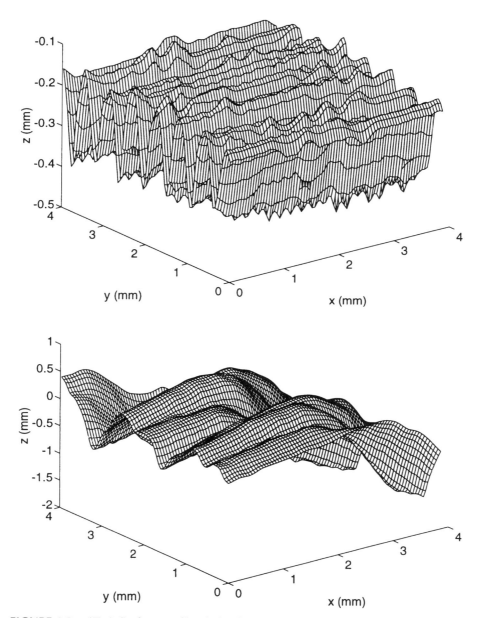

FIGURE 1.3a. (*Top*) Surface profile of aluminum disc in Figure 1.2*a*. (*Bottom*) Surface profile of aluminum disc in Figure 1.2*b*.

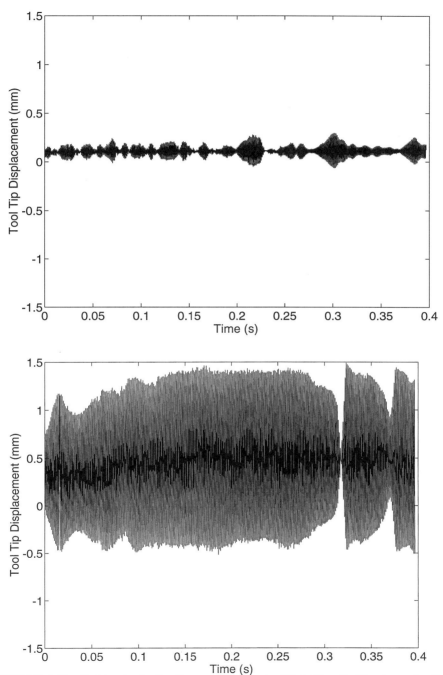

FIGURE 1.3b. Cutting tool vibration measurements. (*Top*) Chaotic-like dynamics for shallow cut in Figure 1.2*a*. (*Bottom*) Periodic-like dynamics for deeper cut in Figure 1.2*b*. (Time scale compressed so that only the envelope is shown.)

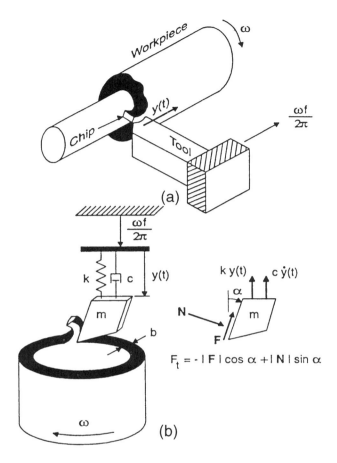

$$F_t = -|F|\cos\alpha + |N|\sin\alpha$$

FIGURE 1.4. Illustration of geometry for turning and simple regenerative chatter model.

with constant angular velocity ω, and the tool is translated along the axis of the workpiece with constant velocity $\omega f/2\pi$. The feed rate f is conventionally given in units of length per revolution, and corresponds to the nominal thickness of the chip removed. The tool generates a surface as material is removed, shown as shaded in Figure 1.4, and any vibrations of the tool are reflected in this surface. In regenerative chatter the surface generated by the tool on one pass becomes the upper surface of the chip on the subsequent pass. The thickness of the chip depends on the present location of the tool and on the location of the tool one revolution ago. Thus differential equations with delay serve as convenient models. In the example illustrated in Figure 1.4, the delay is the time required for the workpiece to complete one revolution.

The experiments of Hooke and Tobias (1963) provide strong evidence for the significance of this regenerative effect. Portions their experimental results are illustrated in Figure 1.5. In Figure 1.5 a hammer blow delivered to the tool results in some tool oscillation, and its corresponding legacy on the cut surface. One revolution later, as shown in Figure 1.5, the legacy of the initial hammer blow causes the tool to vibrate once again. This vibration produces yet more waves on the cut surface. The result is an increase in the amplitude of tool vibration with each successive rotation of the workpiece. As the chatter becomes more fully developed, more and more of the surface becomes wavy, and the once per revolution characteristic of the chatter is less pronounced.

A classical paradigm for regenerative chatter is given by the lumped parameter model in Figure 1.4b. The workpiece in Figure 1.4b rotates with a constant angular velocity ω, and the base of the tool translates at a speed of $\omega f / 2\pi$, where f is the *feed* per revolution. The simplified model in Figure 1.4b represents *orthogonal turning*, since the tool geometry can be represented in two dimensions and is assumed to be constant across the chip width b. The stress distribution on the active face of the tool can be resolved into a normal force $\mathbf{N}$ and a friction force $\mathbf{F}$ on the face of the tool, as shown in Figure 1.4b. The resulting force on the tool in the y-direction, known as the thrust force F_t, is given by $F_t = -|\mathbf{F}| \cos \alpha + |\mathbf{N}| \sin \alpha$, where α is the rake angle of the tool. The equation of motion for the lumped parameter model in Figure 1.4b is then

$$m\ddot{y}(t) + c\dot{y}(t) + ky(t) = -F_t(f + y(t) - y(t - \tau)), \tag{1}$$

where $F_t(\cdot)$ is the thrust force, and this force is a function of the instantaneous chip thickness. The delay τ in (1) is the time required for one revolution of the workpiece $\tau = 2\pi/\omega$. If the tool does not vibrate, the chip thickness is equal to the nominal feed rate f. If (1) is rewritten about the static equilibrium position $\bar{y} = -F_t(f)/k$ by introducing

$$x(t) \equiv y(t) - \bar{y} = y(t) + \frac{F_t(f)}{k}, \tag{2}$$

then (1) becomes

$$\ddot{x}(t) + \frac{c}{m}\dot{x}(t) + \frac{k}{m}x(t) = -\frac{1}{m}[F_t(f + x(t) - x(t - \tau)) - F_t(f)]. \tag{3}$$

Even when the thrust force $F_t(\cdot)$ depends nonlinearly on the chip thickness, the change in chip thickness $x(t) - x(t - \tau)$ is usually small compared to the nominal thickness f. These observations provide some motivation for expanding the function $F_t(\cdot)$ in a Taylor series about f and keeping only the linear

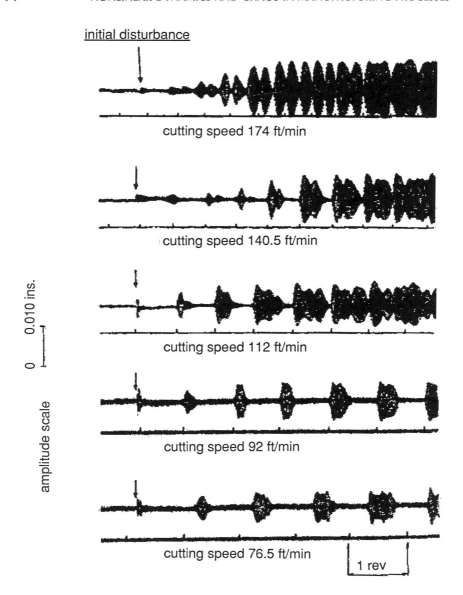

initial disturbance

cutting speed 174 ft/min

cutting speed 140.5 ft/min

cutting speed 112 ft/min

cutting speed 92 ft/min

cutting speed 76.5 ft/min

1 rev

0.010 ins.

0

amplitude scale

FIGURE 1.5. Evidence for regenerative theory of chatter from Hooke and Tobias (1963). Buildup of chatter occurs after a hammer blow in the cutting of mild steel. The tick marks on the horizontal axis represent one revolution of the workpiece.

terms,

$$\ddot{x}(t) + 2\zeta\omega_0\dot{x}(t) + \omega_0^2 x(t) = -\frac{k_s(f)b}{m}\,(x(t) - x(t - \tau)), \tag{4}$$

where

$$\omega_0^2 \equiv \frac{k}{m}, \quad \zeta \equiv \frac{c}{2\sqrt{km}}, \quad \text{and} \quad k_s(f) \equiv \frac{F_t'(f)}{b}, \tag{5}$$

and b is the nominal chip width. Experimentally one finds that $F_t(\cdot)$ is approximately proportional to the chip width, and hence $k_s(f)$ is nearly independent of the chip width. The thrust force $F_t(f)$ is often measured quasi-statically as a function of the nominal feed rate f. Tobias (1965) illustrates that the thrust force $F_t(f)$, as measured using a vibrating tool, can exhibit a hysteretic dependence on a dynamically varying chip thickness f. Thus the quasi-static characterization of the thrust force $F_t(\cdot)$ does not necessarily generalize to the dynamic case. However, the dynamic model in (4) is based on the generalization of quasi-static measurements of $F_t(\cdot)$.

The asymptotic stability of the constant solution of $\bar{y} = -F_t(f)/k$ of (1) is now given by the asymptotic stability of the trivial solution $x(t) = 0$ of (4). The characteristic equation associated with (4) can be determined from Laplace transform techniques or the more general methods presented by Stépán (1989) or Hale (1977). As with linear constant coefficient ordinary differential equations, a necessary and sufficient condition for the asymptotic stability of the trivial solution of (4) is that all roots of the characteristic equation have negative real part. The characteristic equation associated with (4) is

$$s^2 + 2\zeta\omega_0 s + \omega_0^2 = -\frac{k_s(f)b}{m}\,(1 - \exp(-s\tau)). \tag{6}$$

The transcendental nature of (6) makes determination of the roots s difficult. The properties of the roots of characteristic equations similar to that in (6) is the primary focus of the book by Stépán (1989). The method of *D-partitions* has been used by El'sgol'ts and Norkin (1973) and Stépán to characterize the roots of such transcendental characteristic equations. Other techniques, such as the τ-decomposition method (Hsu, 1970; Lee and Hsu, 1969), and adaptations of the Nyquist criterion, can be used as well. Standard root locus techniques used in control theory can also be applied to determine the location of the roots of (6) in the complex plane as a single parameter is varied. A nice summary and analysis of (6) is given by Stépán (1989). (See also this book, Chapter 6.)

As an example using the method of D-partitions, one assumes a purely imaginary root $s = iq$ of (6) with q a real number and substitutes $s = iq$ in (6).

The resulting complex algebraic equations can be solved for $k_s b/m$ and $1/\tau$ parametrically in terms of q as

$$N(q) \equiv \frac{60}{\tau(q)} = 60 \frac{q}{2 \tan^{-1}[(\omega_0^2 - q^2)/(2\zeta\omega_0 q)]},$$

$$\frac{k_s b}{m}(q) = \frac{q^2 - \omega_0^2}{2} + \frac{(2\zeta\omega_0 q)^2}{2(q^2 - \omega_0^2)}, \qquad (7)$$

where N represents the workpiece angular velocity in revolutions per minute. The curves given by the above expression are illustrated in Figure 1.6. The region denoted as stable indicates that all roots have negative real parts in that parameter regime. The diagram in Figure 1.6 will be referred to as a *stability diagram* for (4).

The stability diagram shown in Figure 1.6 is an oft-cited diagram for regenerative chatter. Diagrams similar to that in Figure 1.6 are constructed by Tobias for processes including milling and drilling. The lobed structure of

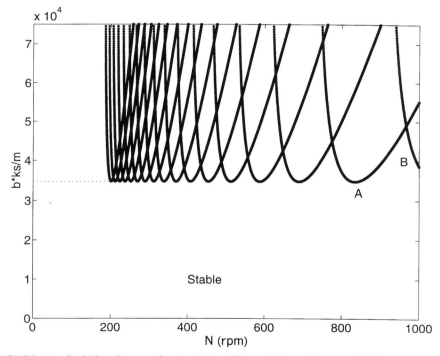

FIGURE 1.6. Stability diagram for the delay-differential equation model of regenerative chatter, Equation (4) ($\omega_0 = 377.0$, $\rho = 0.11$). Vertical axis is proportional to the cutting width and depth.

stability diagrams has been used by Smith and Delio to increase material removal rates by increasing the chip width b. If one encounters chatter at a given chip width b, say, at point A in Figure 1.6, this regenerative chatter may be avoided by changing the speed, say, to that at point B. At point B the trivial solution of (4) is asymptotically stable, regenerative chatter does not occur, and the production rate has been improved by taking larger cuts at higher speeds than possible at point A.

The simplicity of (4), and its widespread use as an explanation for regenerative chatter, lead to labeling (4) as the *classical model for regenerative chatter*. The minimum value of $k_s b/m$ along the lobes in Figure 1.6 is

$$\left[\frac{k_s b}{m}\right]_{\min} = 2\zeta\omega_0^2(1 + \zeta)$$

which can be demonstrated by finding the extrema of the expression for $k_s b/m(q)$ in (7). Thus, regardless of the cutting speed N, the classic model of regenerative chatter indicates that no tool vibrations can occur below this minimum threshold. For a given value of k_s/m, the classic model indicates that tool vibration cannot occur for small chip widths. Despite this prediction, small amplitude tool vibrations can occur well below the threshold chip thickness predicted by the classic model of regenerative chatter.

1.7   NONLINEAR MODELS

The use of linear differential difference equations has dominated the research literature on machine tool vibrations and chatter. The book by Tobias provides numerous other examples of linear constant coefficient differential difference equations used as models for chatter in milling, drilling, and grinding, as well as turning.

Despite the dominance of linear constant coefficient differential difference equations in the literature on machine tool vibrations, the potential importance of nonlinearities in describing tool chatter has been known for some time, though perhaps not studied as extensively. Arnold (1946), whose paper is one of the earliest works devoted specifically to tool chatter, proposed the nonlinear differential equation

$$M\ddot{x}(t) - [A + Bx - \phi(x)]\dot{x}(t) + F'x(t) = K, \tag{8}$$

where M, A, B, F', and K are constants, and $\phi(x)$ is an unspecified function that represents the dependence of the system damping on the tool motion $x(t)$. An early nonlinear model that accounts for the separation of the tool and workpiece is that given by Doi and Kato (1956).

Hanna and Tobias (1974) introduced the nonlinear differential difference equation

$$\frac{\ddot{x}(t)}{p^2} + \frac{b}{\lambda}\frac{\dot{x}(t)}{\omega} + x(t) + \beta_1 x^2(t) + \beta_2 x^3(t)$$

$$= -\frac{k_1}{\lambda}\left[(x(t) - x(t - \tau)) + C_1(x(t) - x(t - \tau))^2 + C_2(x(t) - x(t - \tau))^3\right]$$

$$(9)$$

to describe regenerative chatter in milling. The constants p, b, λ, ω, β_1, β_2, k_1, C_1, and C_2 all relate to a lumped parameter one degree of freedom model. Lin and Weng (1991) developed the two degrees of freedom model for tool vibrations in turning,

$$\ddot{x}(t) + 2\zeta\dot{x}(t) + x(t) = -dF_x,$$
$$\ddot{y}(t) + 2l_1\zeta\dot{y}(t) + l_2 y(t) = -dF_y,$$

$$(10)$$

where dF_x, dF_y are cubic polynomials in $x(t)$, $x(t - \tau)$, $y(t)$, $y(t - \tau)$. Lin and Weng's model also includes the potential for the tool to leave the workpiece.

Berger, Rokni, and Minis (1992) (see also this book, Chapter 5) also considered a system with two degrees of freedom, denoted by $x_1(t)$ and $x_2(t)$, as a model for vibrations in turning;

$$\dot{x}_1(t) = v_1(t),$$
$$\dot{x}_2(t) = v_2(t),$$
$$\dot{v}_1(t) = -a_8 x_1(t) - a_7 v_1(t) + a_1 x_1(t - \tau) + (x_1(t) - x_1(t - \tau))$$
$$[a_3(v_2(t) - v_2(t - \tau)) - a_2(v_1(t) - v_1(t - \tau))] - a_{11}v_1^3(t),$$
$$\dot{v}_2(t) = -a_4 x_1(t) - a_{10}x_2(t) - a_9 v_2(t) - a_4 x_1(t - \tau) + (x_1(t) - x_1(t - \tau))$$
$$[a_6(v_2(t) - v_2(t - \tau)) - a_5(v_1(t) - v_1(t - \tau))] - a_{12}v_2^3(t),$$

$$(11)$$

where the twelve constants a_k for $k = 1, 2, 3, \ldots, 12$, depend on the particular cutting conditions, and τ is the time required for one revolution of the workpiece.

1.8   FRICTIONLIKE MODELS

Several investigators have included friction effects between the workpiece and the tool, and the chip and tool, in models for machine tool vibrations. Others have included cutting force models that mimic characteristics seen in frictional oscillators. Arnold's experiments showed that the cutting force decreased with increasing cutting speed, as shown in Figure 1.7. The decline in the cutting force with relative veloity is analogous to that seen in dry friction. The nonlinear dynamics of friction oscillations are discussed in this book by Shaw and Feeny, Chapter 9.

A relatively simple one degree of freedom model derived by Hamdan and Bayoumi (1989), which supposedly incorporates friction effects on the rake and flank faces of the tool, is

$$m\ddot{y}(t) + \left[C - C_e \left(\frac{1}{1 - \dot{y}(t)/v_0} \right) \right] \dot{y}(t) + (k - k_e)y(t) = 0, \qquad (12)$$

where m, C, C_e, k, k_e are constants and v_0 is the constant velocity with which the workpiece moves past the tool.

A frictionlike model with two degrees of freedom has been proposed by Grabec (1986, 1988). This model provides some theoretical foundation for claims that machine tool vibrations can be chaotic. Grabec developed the two

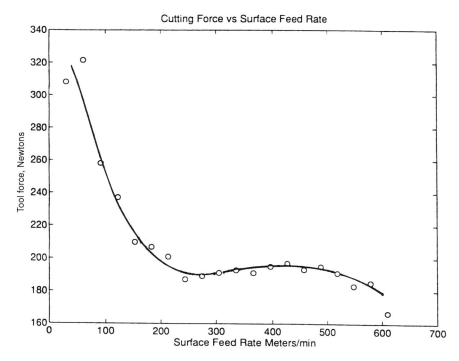

FIGURE 1.7. Cutting force versus speed for orthogonal cutting, 7075 aluminum (Hardinge CNC machine, Cornell University).

degrees of freedom model illustrated in Figure 1.8. The differential equations developed by Grabec are

$$m\ddot{x}(t) + c_x\dot{x}(t) + k_x x(t) = F_{x0}\left(\frac{h_0 - y(t)}{\bar{h}_0}\right)\left[C_1\left(\frac{v_0 - \dot{x}(t)}{\bar{v}_0} - 1\right)^2 + 1\right]$$

$$\times \mathcal{H}\left(\frac{h_0 - y(t)}{\bar{h}_0}\right)\mathcal{H}\left(\frac{v_0 - \dot{x}(t)}{\bar{v}_0}\right),$$

$$m\ddot{y}(t) + c_y\dot{y}(t) + k_y y(t) = K_0 F_{x0}\left(\frac{h_0 - y(t)}{\bar{h}_0}\right)\left[C_1\left(\frac{v_0 - \dot{x}(t)}{\bar{v}_0} - 1\right)^2 + 1\right]$$

$$\times \left[C_2\left(\frac{v_0 - \dot{x}(t) - R\dot{y}(t)}{\bar{v}_0} - 1\right)^2 + 1\right]$$

$$\times \left[C_3\left(\frac{h_0 - y(t)}{\bar{h}_0} - 1\right)\right]\text{sign}\left(\frac{v_0 - \dot{x}(t) - R\dot{y}(t)}{\bar{v}_0}\right)$$

$$\times \mathcal{H}\left(\frac{h_0 - y(t)}{\bar{h}_0}\right)\mathcal{H}\left(\frac{v_0 - \dot{x}(t)}{\bar{v}_0}\right),$$

$$R = R_0\left[C_4\left(\frac{v_0 - \dot{x}(t)}{\bar{v}_0} - 1\right)^2 + 1\right], \tag{13}$$

where $\bar{v}_0$, $\bar{h}_0$, C_1, C_2, C_3, and C_4 are treated as constants, and $\mathcal{H}(\cdot)$ is the unit step function. The two step functions $\mathcal{H}$ in (13) account for the possibility that the tool can loose contact with the workpiece, and the sign($\cdot$) function allows the y-component of the friction force to change directions. However, unlike the curves illustrated in Figure 1.7, Grabec's choice of friction characteristic [terms with C_1 in (13)] begins to increase with velocity after an initial decrease instead of monotonically decreasing or reaching some fixed minimum.

The significance of the friction between the chip and the tool, or cutting forces that have the frictionlike characteristics shown in Figure 1.7, in main-

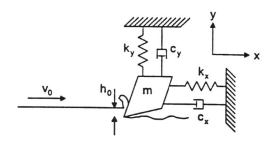

FIGURE 1.8. Two-degree-of-freedom chaotic dynamics model of Grabec (1986) for cutting of metals.

taining tool chatter has yet to be resolved. It is clear from work on friction oscillators that systems with two or more degrees of freedom are necessary to explain aperiodic tool vibrations. The two degrees of freedom model used by Grabec, and the four-dimensional phase space associated with this system, apparently give rise to chaotic oscillations. A discussion of chaos in machining is given below.

1.9   PERIODIC CHIP FORMATION AND SHEAR BANDING

Another potential mechanism for chatter is the excitation of the machine tool structure by periodic shearing actions in the chip as material is removed. In principle, even a rigid tool would experience periodic forcing due to the nature of chip formation. Landberg (1956) conducted an investigation of the periodic patterns in segmented chips. Based on the spatial distribution of chip features, along with a knowledge of the cutting speed, Landberg calculated a temporal frequency associated with chip formation. Landberg claimed to have found a resonance phenomenon when the chip formation frequency is close to a natural frequency of the tool. Albrecht (1961) partially attributes fatigue cracks on tool flanks to this periodic shearing action in the chip formation, and the resultant cyclic loading on the tool. Doi and Ohhashi (1992) performed experiments that seem to indicate that the frequencies associated with chip formation are almost directly proportional to the cutting speed.

Davies, Chou, and Evans (1996) (see also this book, Chapter 3), have reported segmented chip formation in the turning of hard steels using Cubic Boron Nitride (CBN) cutting tools. The segmented chips reported by Davies, Chou, and Evans have periodic features at the micron level ($\approx 30 \times 10^{-6}$ meters) at cutting speeds of 1–5 meters per second. The corresponding temporal frequency is > 50 kHz, while typical tool frequencies are 1–2 kHz. The frequencies measured by Landberg, Albrecht, and Doi and Ohhashi are apparently much lower than the spatial and temporal frequencies reported by Davies, Chou, and Evans.

The possibility that tool vibrations are caused by interactions between material instabilities in the chip formation process and the machine tool structure is intriguing. A potential material instability is fracture in the chip zone. There is growing evidence that dynamic fracture may be a chaotic process (see Marder, Chapter 10). Another fundamental material instability is the elastoplastic thermal instability in metals called "shear banding" studied by Wright (1990). But direct application of fracture mechanics and shear banding to machining dynamics has not been made.

1.10   EVIDENCE FOR NONLINEAR DYNAMICS IN MACHINING

A large body of experimental work has been devoted to characterizing and quantifying the dynamics of the cutting process. The review by Tlusty (1978) provides an overview and a large number of references to such works. An earlier work that claims evidence for the force delay model is the work of Doi and Kato (1956).

Hooke and Tobias (1963) developed a set of novel experiments in which they deliver a hammer blow to a tool while turning mild steel. They showed that the hammer blow can induce chatter under cutting conditions that are normally chatter free. They generated a diagram, shown in Figure 1.9, familiar to those acquainted with the Hopf bifurcation in nonlinear dynamics. The points along the curves labeled A in Figure 1.9 represent experiments without the hammer blow, while those along the curves labeled B denote experiments with the hammer blow. From Figure 1.9a it appears that for a width of cut greater than 0.01 inches, the blow caused the system to vibrate with a finite

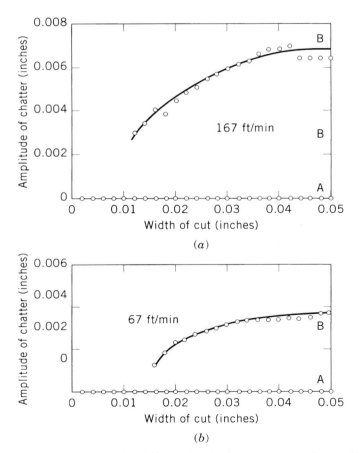

FIGURE 1.9. Evidence for subcritical bifurcation in chatter dynamics for cutting metals from Hooke and Tobias (1963): (a) 167 ft/min, 0.00185 inch/rev; (b) 67 ft/min, 0.00185 inch/rev.

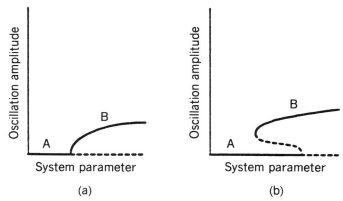

FIGURE 1.10. Sketch of supercritical (*a*) and subcritical (*b*) Hopf bifurcation. Solid lines represent stable branches, dotted lines represent unstable branches. System parameter in Figure 1.9 is the width of cut.

amplitude. Figure 1.9 might be interpreted as a *subcritical bifurcation* like that shown in Figure 1.10. In this case the trivial solution would remain stable for widths of cut larger than 0.01 inches in *a* and 0.015 inches in *b*, and the hammer blow is sufficient to move the system onto the stable branch B in Figure 1.10.

Ten years later Hanna and Tobias (1974) revisited the experiments described by Hooke and Tobias. Hanna and Tobias performed a series of experiments in milling which are analogous to the turning experiments of Hooke and Tobias.

Subcritical bifurcation instabilities are the most difficult and *sometimes dangerous* nonlinear phenomena to deal with because one cannot use linear theory to predict them. Also small random events can cause the system to jump from small to large amplitude vibrations (from branch A to B in Figure 1.10*b*). Several new studies of subcritical Hopf bifurcations in machining have appeared (see Nayfeh et al. and Stépán, Chapters 6 and 7 in this book, and Johnson, 1996). A later work by Tobias on nonlinear theory of chatter is that of Shi and Tobias (1984).

1.11   IS CHAOS DESIRABLE IN MACHINING?

With the speculation of Grabec (1986) that machining dynamics might be chaotic, several research groups have conducted experiments to test this hypothesis. Moon (1994), Moon and Abarbanel (1995), and Bukkapatnam, Lakhtakia, and Kumara (1995) have provided experimental evidence that tool vibrations in turning can be chaotic. Moon used reconstructed phase portraits, probability distribution functions, and Fourier power spectra to support the notion that tool vibrations in the turning of aluminum are chaotic. Bukkapatnam, Lakhtakia, and Kumara (1995) applied statistical measures and calculated Lyapunov exponents to verify that dynamic force measurements in the turning steel exhibit low-dimensional chaotic behavior. Berger et al. (1995) used the method of false nearest neighbors (FNN) to show that tool accelerations in the turning of mild steel exhibit low-dimensional chaos. Berger et al. further assert that the chaotic attractor can be embedded in a four-dimensional space, and thus has a dimension $\leqslant 4$.

An example of the evidence for chaos in machining below the chatter regime is shown in Figures 1.11 and 1.12. These data and analysis are from the work of Moon and Abarbanel (1995). The sample is the same aluminum disc as shown in Figures 1.1 and 1.2. The depth of cuts ranged from 0.025 mm to 0.127 mm which were believed to be below the chatter threshold. The method of *false nearest neighbors* (FNN) as outlined in Abarbanel (1996) is used. The method uses a time series data measurement, say, $\{x_0, x_1, x_2, \ldots, x_{k-1}, x_k, x_{k+1}, \ldots\}$ to construct a pseudo phase space of n-dimensional vectors $[x_k, x_{k+1}, \ldots, x_{k+n-1}]^T$. If the dimension n is chosen too low, then the orbit will appear to cross itself, as when a three-dimensional orbit is projected onto

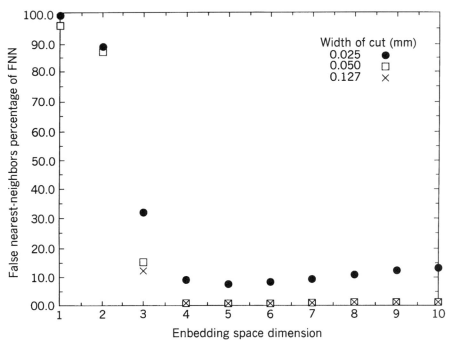

FIGURE 1.11. Evidence for chaotic dynamics in cutting aluminum (see Figures 1.2 and 1.3). The number of false crossing (neighbors) in the phase space falls to near zero for dimension 4 or greater for cuts of 0.050 and 0.127 mm. The solid dots show randomness at a very small cut. (From unpublished paper of Moon and Abarbanel, 1995.)

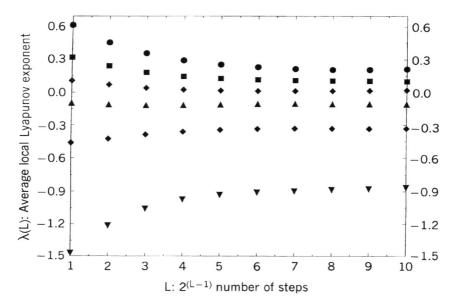

FIGURE 1.12. Evidence for chaotic dynamics in cutting aluminum (see Figures 1.2 and 1.3). Two positive Lyapunov exponents calculated from time series dynamics of the cutting tool indicate divergence of trajectories in the phase space.

a plane. The basic idea of FNN is to create a large enough phase space so that the orbit is unraveled and does not cross itself.

The algorithm created by the San Diego group of Abarbanel plots the percentage of FNN due to projection orbit crossing. As expected, as one increases the size of the embedding space, the number of FNN → 0.

In the case of machining dynamics in Figure 1.11, the phase space appears to be below $n = 5$. This suggests that the randomlike dynamics are in fact generated by a low-dimensional set of differential equations. To test for chaos, we employed another UCSD algorithm developed by the Abarbanel group to test for positive Lyapunov exponents. The results are shown in Figure 1.12 for the machining case. The two positive exponents imply that the small random-like dynamics in the cutting is indeed chaotic. A positive Lyapunov exponent measures the rate at which nearby trajectories in the phase space are diverging on average. (See Moon, 1992 for an elementary discussion of Lyapunov exponents and chaos.) Johnson (1996) has performed a set of experiments, motivated by the Hooke and Tobias experiments, that were designed to produce a qualitative picture of the low-dimensional chaotic attractor suggested by the above-mentioned chaos experiments.

The experimental evidence suggestive of a subcritical Hopf bifurcation in Figure 1.9, and the mounting evidence that tool vibrations can be chaotic, emphasizes the significance of nonlinear dynamics in describing tool vibrations and chatter. Moreover the strong experimental evidence for the regenerative effect presented in Figure 1.5 suggests that nonlinear differential difference equations are required to describe the physics of the cutting process. The existence of small amplitude vibrations before the onset of chatter highlights a key limitation in the classic model of regenerative chatter.

1.12   NONREGENERATIVE CHATTER AND CHAOS

The classical model for chatter suffers from the fact that it does not predict any dynamics below the instability cutting depth threshold, nor does it predict nonlinear phenomena such as chaos and subcritical bifurcations. Grabec recognized that chaotic dynamics demanded a phase space dimension greater than two and constructed a nonlinear two-degree-of-freedom nondelay friction model for the dynamics. However, another possibility lies in an earlier model for chatter proposed in Japan by Doi and Kato (1956). They performed cutting experiments and claimed to have measured a time delay in the cutting force that was much smaller than the workpiece rotation period. Such a short delay time could be related to shear banding instabilities in the cutting zone or to the time for the chip to move across the cutting tool. Recently in a dissertation at Cornell University, Johnson (1996) has explored the dynamic bifurcation of a model based on the Doi and Kato model. Johnson, however, used a different nonlinear term and analyzed a nondimensional equation of the form

$$\ddot{x}(t) + \gamma_1 \dot{x}(t) + \gamma_2[x(t) + x^3(t)] = -\gamma_3 x(t-1). \tag{14}$$

The coefficients γ_1, γ_2 reflect the tool damping and stiffness, respectively. The constant γ_3 is proportional to the width of the cut. Time is nondimensionalized by a delay time, which Doi and Kato measured approximately as 0.5 ms. Note that the time delay of a workpiece rotating at 1200 rpm is 50 ms.

In the modified Doi-Kato model above, the vibration amplitude is non-dimensionalized by a cubic nonlinearity in the tool stiffness. A Poincaré map has been defined by plotting the values of position and velocity, $\{x_n, \dot{x}_n\}$ when the delayed variable reaches a certain level, namely $x(t-1) = x_0$, and $\dot{x}(t-1) > 0$. Numerical simulation of (14) using the values $\gamma_1 = 0.0415$, $\gamma_2 = 0.0355$ was carried out for the range of cutting force parameters $0.05 \leqslant \gamma_3 \leqslant 38.0$. The resulting Poincaré maps are shown in Figure 1.13. The single dots in Figure 1.13a indicate periodic limit cycle dynamics, whereas the closed loop in Figure 1.13b–c show a quasi-periodic motion. These Poincaré maps give evidence for chaotic motion as the quasi-periodic torus develops fractal-looking wrinkles. These pictures show the complexity of motion possible in the simple nonlinear differential equation with time delay.

Although the values of $\gamma_1, \gamma_2, \gamma_3$ were chosen to be representative of cutting tool properties, the numerical simulation shows an increase in the carrier frequency as γ_3 is increased, which is not observed in our experiments. A

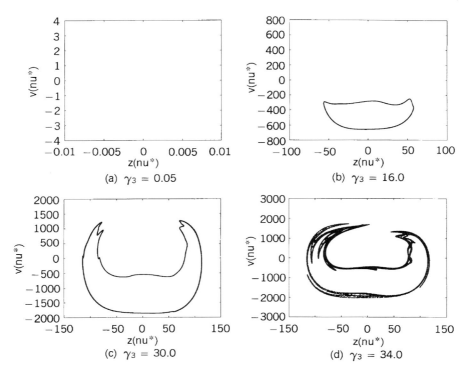

FIGURE 1.13. Poincaré section of dynamics from a mathematical model of regenerative chatter with time delay. A dot indicates periodic motion, closed curve represents quasi-periodic motion, fractal-like dynamics indicate possible chaos. (From Johnson, 1996.)

combined short delay–regenerative delay model has also been studied by Johnson which shows simulation dynamics closer to actual experiments:

$$\ddot{x} + 2\tau\omega_0\dot{x} + \omega_0^2 x = -\kappa[x(t - h) - x(t - h - \tau)], \tag{15}$$

where h is the short delay and τ is the regenerative delay equal to the rotation period. Johnson (1996) has shown that a model with both short and regenerative delays results in a critical value of the cutting depth less than half of that predicted by the classical regenerative model. Results of the analysis of this model are to be reported in a future research journal paper.

1.13 SUMMARY

The above discussion on machine tool chatter illustrates some of the complexities that nonlinearities can introduce in material processing dynamics as well as some of the short-comings of the simple models presented. Differential-delay models are easily amenable to modern analytic and numerical tools that can show which parameters are most sensitive in producing chatter behavior. These simple models are aso useful for designing control systems to suppress chatter. However, they clearly fail to capture potentially important features of cutting dynamics, such as material and chip instabilities. This is illustrated in the finite-element simulations shown in Figure 1.14, which were performed by Professor David Benson of the University of California, San Diego. The data were for high-strength steel at a very high cutting speed. The sequence of pictures shows the creation of a chip and its subsequent unstable deformation even when the tool has infinite stiffness. Not shown in this figure are the temperature dynamics, which are neglected in the classical chatter models.

These observations suggest the following challenges and opportunities for the application of nonlinear dynamics to manufacturing processes:

- How do chaotic and random dynamics during processing affect surface and overall product quality?

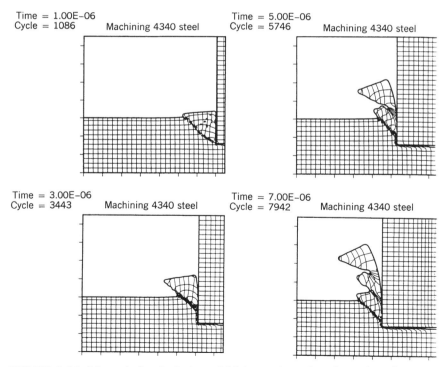

FIGURE 1.14. Numerical calculation of high-speed cutting dynamics of 4340 steel showing material instability of the chip for a rigid tool motion. (Courtesy of Professor David Benson, University of California, San Diego.)

- Can nonlinear dynamics be used for new diagnostic techniques to predict tool damage and monitor and control overall process quality?
- How can local material dynamics, including temperature, fracture, and material instabilities, be incorporated in low-dimensional, intuitively appealing, and conceptually simple dynamic models?
- Can low-dimensional models be used to design chatter avoidance and dynamics enhancement control systems for manufacturing processes?

This book came out of a workshop sponsored by the Institute of Mechanics and Materials in March 1995. The workshop was attended by 80 participants with representatives from 8 industrial companies. As a result of round-table discussion at the end of this workshop, a consensus emerged as to the top priority research topics in manufacturing process dynamics. We end this chapter with a list of these topics which we hope will serve as a benchmark for measuring progress in this field decades hence.

Workshop Consensus Issues

- The control of chatter in metal cutting remains an important issue especially in the aerospace industry.
- The control of chatter and dynamics in the rolling of metals remains a continuing problem in industry.
- Better models for cutting physics are needed to control unwanted dynamics.
- Chaotic dynamics may be normal in pre-chatter cutting of materials.
- New control strategies based on chaos theory may have application in material processing.
- To achieve practical application of nonlinear dynamics in manufacturing processes, researchers must eventually work with machine tool suppliers.
- New nonlinear dynamics concepts may lead to new diagnostic methods in material processing such as predictions of tool wear and surface quality.

REFERENCES

Abarbanel, H. 1996. *Analysis of Observed Chaotic Data.* Springer-Verlag, New York (in press).

Ashley, S. 1995. High-speed machining goes mainstream. *Mech. Eng.:* 56–61.

Albrecht, P. 1961. New developments in the theory of the metal-cutting process: Part II, The theory of chip formation. *Trans. ASME J. Eng. Ind.* **83**:557–571.

Arnold, R. N. 1946. The mechanism of tool vibration in the cutting of steel. *Proc. Inst. Mech. Eng.* **154**:261–284.

Berger, B. S., Minis, I., Chen, Y. H., Chavali, A., and Rokni, M. 1995. Attractor embedding in metal cutting. *J. Sound Vib.* **18**:936–942.

Berger, B. S., Rokni, M., and Minis, I. 1992. The nonlinear dynamics of metal cutting. *Int. J. Eng. Sci.* **30**:1433–1440.

Bukkapatnam, S. T. S., Lakhntakia, A., and Kumara, S. R. T. 1995. Analysis of sensor signals shows turning on a lathe exhibits low-dimensional chaos. *Phys. Rev. E* **52**:2375–2387.

Comstock, T. R., Tse, F. S., and Lemon, J. R. 1968. Chatter suppression by controlled mechanical impedance. American Society of Tool and Manufacturing Engineers, Paper Number MR68-101, pp. 1–12.

Davies, M. A., Chou, Y. S., and Evans, C. J. Evans. 1996. On chip morphology tool wear and cutting mechanics in finish hard turning. *Ann. CIRP* **45**.

Delio, T., Tlusty, J., and Smith, S. 1992. Use of audio signals for chatter detection and control. *Trans. ASME J. Eng. Ind.* **114**:146–157.

Doi, M., and Ohhashi, M. 1992. A study on parametric vibration in machining of hard cutting metals. *Int. J. Jap. Soc. Precision Eng.* **26**:195–200.

Doi, S., and Kato, S. 1956. Chatter vibration of lathe tools. *Trans. ASME* **78**:1127–1134.

El'sgol'ts, L. E., and Norkin, S. B. 1973. *Introduction to the Theory and Application of Differential Equations with Deviating Arguments.* Academic Press, San Diego.

Grabec, I. 1986. Chaos generated by the cutting process. *Phys. Lett. A* **117**:384–386.

Grabec, I. 1988. Chaotic dynamics of the cutting process. *Int. J. Mach. Tools Manuf.* **28**:19–32.

Hahn, R. S. 1951. Design of lanchester damper for elimination of metal-cutting chatter. *Trans. ASME* **73**:331–335.

Hahn, R. S. 1953. Metal-cutting chatter and its elimination. *Trans. ASME* **75**:1073–1080.

Hale, J. K. 1977. *Theory of Functional Differential Equations. Applied Mathematical Sciences*, vol. 3. Springer-Verlag, New York.

Hamdan, M. N., and Bayoumi, A. E. 1989. An approach to study the effects of tool geometry on the primary chatter vibration in orthogonal cutting. *J. Sound Vib.* **128**:451–469.

Hanna, N. H., and Tobias, S. A. 1974. A theory of nonlinear regenerative chatter. *Trans. ASME J. Eng. Ind.* **96**:247–255.

Hooke, C. J., and Tobias, S. A. 1963. Finite amplitude instability—A new type of chatter. *Proc. Fourth Int. MTDR Conf.* Pergamon Press, Manchester, pp. 97–109.

Hsu, C. S. 1970. Application of the tau-decomposition method to dynamical systems subjected to retarded follower forces. *Trans. ASME J. Appl. Mech.* **37**:259–266.

Johnson, M. A. 1966. Nonlinear differential equations with delay as models for vibrations in the machining of metals. Ph.D. dissertation. Cornell University, Ithaca.

Klein, R. G., and Nachtigal, C. L. 1975a. The application of active control to improve boring bar performance. *Trans. ASME J. Dynamic Systs. Meas. Control* **97**:179–183.

Klein, R. G., and Nachtigal, C. L. 1975b. A theoretical basis for the active control of a boring bar operation. *Trans. ASME J. Dynamic Systs. Meas. Control* **97**:172–178.

Landberg, P. 1956. Vibrations caused by chip formation. *Microtenic* **10**:219–228.

Lee, M. S., and Hsu, C. S. 1969. On the tau-decomposition method of stability analysis for retarded dynamical systems. *Soc. Ind. Appl. Math. (SIAM) J. Control* 7:242–259.

Lemon, J. R., and Long, G. W. 1965. Survey of chatter research at the Cincinnati Milling Machine Company. In Tobias, S. A., and Koenigsberger, F., eds., *Advances in Machine Tool Design and Research. Proc. Fifth Int. MTDR Conf.*, Pergamon Press, Manchester, pp. 545–586.

Lin, J. S., and Weng, C. I. 1991. Nonlinear dynamics of the cutting process. *Int. J. Mech. Sci.* 33:645–657.

Lin, S. C., DeVor, R. E., and Kapoor, S. G. 1990. The effects of variable speed cutting on vibration control in face milling. *Trans. ASME J. Eng. Ind.* 112:1–11.

Marusich, T. D., and Ortiz, M. 1995. Modelling and simulation of high-speed machining. *Int. J. Num. Methods Eng.* 38:3675–3694.

Merritt, H. E. 1965. Theory of self-excited machine-tool chatter. *Trans. ASME J. Eng. Ind.* 87:447–454.

Moon, F. C. 1992. *Chaotic and Fractal Dynamics.* Wiley, New York.

Moon, F. C. 1994. Chaotic dynamics and fractals in material removal processes. In Thompson, J. M. T., and Bishop, S. R., eds., *Nonlinearity and Chaos in Engineering Dynamics.* Wiley, New York, pp. 25–37.

Moon, F. C., and Abarbanel, H. 1995. Chaotic motions in normal cutting of metals. Cornell University, Sibley School of Mechanical and Aerospace Engineering Report.

Nachtigal, C. 1972. Design of a force feedback chatter control system. *Trans. ASME J. Dynamic Systs. Meas. Control* 94:5–10.

Nachtigal, C., and Cook, N. H. 1970. Active control of machine-tool chatter. *Trans. ASME J. Basic Eng.* 92:238–244.

Optiz, H., Dregger, E. U., and Roese, H. 1966. Improvement of the dynamic stability of the milling process by irregular tooth pitch. In *Proc. Seventh Int. MTDR Conf.* Pergamon Press, Manchester, pp. 213–227.

Oxley, P. L. B., and Hastings, W. F. 1977. Predicting the strain rate in the zone of intense shear in which the chip is formed in machining from the dynamic flow stress properties of the work material and the cutting conditions. *Proc. R. Soc. London A* 356:395–410.

Saravanja-Fabris, N., and D'Souza, A. F. 1974. Nonlinear stability analysis of chatter in metal cutting. *Trans. ASME J. Eng. Ind.* 96:670–675.

Shi, H. M., and Tobias, S. A. 1984. Theory of finite amplitude machine tool instability. *Int. J. Mach. Tool Des. Res.* 34:45–69.

Sisson, T. R., and Kegg, R. L. 1969. An explanation of low-speed chatter effects. *Trans. ASME J. Eng. Ind.* 97:951–958.

Smith, S., and Tlusty, J. 1990. Update on high-speed milling dynamics (1990). *Trans. ASME J. Eng. Ind.* 112:142–149.

Stépán, G. 1989. *Retarded Dynamical Systems: Stability and Characteristic Functions.* Pitman Research Notes in Mathematics, vol. 210. Longman Scientific and Technical, London.

Taylor, F. W. 1907. On the art of cutting metals. *Trans. ASME* 28:31–350.

Tlusty, J. 1978. Analysis of the state of research in cutting dynamics. *Ann. CIRP* 27:583–589.

Tlusty, J., and Ismail, F. 1981. Basic non-linearity in machining chatter. *CIRP Ann. Manuf. Tech.* **30**:299–304.

Tobias, S. A. 1965. *Machine Tool Vibration.* Blackie and Son, Glasgow.

Ulsoy, A. G., and Koren, Y. 1993. Control of machining processes. *Trans. ASME J. Dynamic Systs. Meas. Control* **115**:301–308.

Welbourn, D. B., and Smith, J. D. 1970. *Machine-Tool Dynamics: An Introduction.* Cambridge University Press, Cambridge.

Wright, T. 1990. Adiabatic shear bands. *Appl. Mech. Rev.* **43**:5196–5200 (Pt. 2, May).

Zhang, G. M., and Kapoor, S. G. 1991. Dynamic generation of machined surfaces. I: Description of a random excitation system. *Trans. ASME J. Eng. Ind.* **113**:137–144.

Zhang, G. M., and Kapoor, S. G. 1991. Dynamic generation of machined surfaces. II: Construction of surface topography. *Trans. ASME J. Eng. Ind.* **113**:145–153.

2

DYNAMIC MODELING AND CONTROL OF MACHINING PROCESSES

A. GALIP ULSOY

2.1 INTRODUCTION

2.1.1 Motivation and Significance

Manufacturing process automation continues to increase in importance over the last few decades due to increased global competition. It has been at, or near, the top of the lists of technologies identified as critical to national economic competitiveness and security in recent studies (Ulsoy and Koren, 1993). A strong manufacturing industry is clearly essential for continued economic development; manufacturing directly accounts for approximately one-quarter of the value of all goods and services produced in industrial nations and has strong links to other sectors of the economy. Its importance for national security is also clear; the conversion of the U.S. manufacturing enterprise to wartime production was essential to victory in World War II.

Metal removal, or machining, processes are widely used in manufacturing and include operations such as turning, milling, drilling, and grinding. The trend toward automation in machining has been driven by the need to maintain high product quality while improving production rates. The potential economic benefits of automation in machining are illustrated schematically in Figure 2.1. Consider, for example, an investment of $30,000 in a controller for

Dynamics and Chaos in Manufacturing Processes, Edited by Francis C. Moon.
ISBN 0-471-15293-5 © 1998 John Wiley & Sons, Inc.

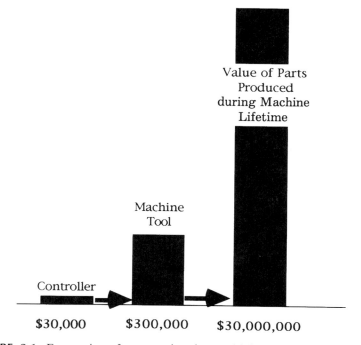

FIGURE 2.1. Economics of automation in machining. A 10% investment in the controller leads to, say, a 10% increase in machine productivity, which can produce a return on investment of 1000 times the original investment.

a $300,000 machine tool that typically produces parts worth $30,000,000 during its lifetime. Assume that this 10% additional investment to increase the controller capability results in an effective increase of 10% in the productivity of the machining operation. This increase in productivity can be as a result of increased metal removal rates, reduced scrap or rework due to improved quality, reduced process downtime, or some combination of these effects. Since during its lifetime the machine produces products valued at $30,000,000, this 10% increase in effective productivity leads to an economic return of $3,000,000. A return on investment of 1000 times ($=3,000,000/3000$) is achieved. It is this two-stage economic amplification, as illustrated in Figure 2.1, that makes the automation of machining processes so critical.

2.1.2 Background

Machining operations represent a mature technology that dates back to the eighteenth century. There have been many improvements in machines, cutting tools, and materials that have led to significant benefits. One of the most important developments has been the introduction of automation since the end of World War II. In the 1950s numerically controlled (NC) machine tools were developed under an Air Force contract to Parson's Machine Tool Company (Traverse City, Michigan) by the Servomechanisms Laboratory at MIT. These reprogrammable but hard-wired digital devices represented the state-of-the-art into the 1960s. During the 1960s and 1970s computer technology and NC machine tools continued to evolve. Computers became not only more powerful but also less expensive and more reliable. The servo-control function (including interpolators for multiaxis coordination) became implemented using on-board computers rather that hard-wired digital circuits. These computer numerically controlled (CNC) systems, because of their powerful computing capabilities, led to advances in the interpolators and in the servo-control loops.

The additional on-board computing power of CNC machines also led to research interest in implementing a higher-level process control aimed at improving production rates and product quality. These process control systems are commonly referred to as "adaptive control" (AC) systems in the manufacturing community, but may or may not include adaptation in the sense in which the term is used in the control literature (Ulsoy, Koren, and Rasmussen, 1983). The first of these process control systems, which was designed under an Air Force contract by Bendix, was an ambitious system that was only demonstrated in the laboratory. It included advanced control features such as on-line optimization and adaptive control and was termed an adaptive control with optimization (ACO) system. However, a major drawback of this system was the need for an accurate on-line measurement or estimate of tool wear. Such measurement methods were not available in the 1960s, and even today a reliable tool wear measurement method that can operate in an *industrial* environment does not exist (Koren and Ulsoy, 1989). Subsequent systems were often less ambitious than the original Bendix concept and were designed to operate on the constraints of the permissible work space (e.g., maximum cutting force). These systems are often classified as adaptive control with constraints (ACC) and geometric adaptive control (GAC) (Ulsoy, Koren, and Rasmussen, 1983; Koren and Ulsoy, 1989). Several commercial devices of the ACC type were also put on the market during the 1970s but did not achieve widespread industrial acceptance (Ulsoy, Koren, and Rasmussen, 1983). Some of the problems that led to the poor performance of these first-generation ACC systems are discussed in a subsequent section. The need for measurement of wear in the Bendix system also generated considerable research interest in on-line tool wear measurement and estimation (Ulsoy and Koren 1989). For a more detailed discussion on the state-of-the-art in machine tool automation through the 1970s, the reader is referred to Koren (1983), Ulsoy, Koren, and Rasmussen (1983), and Ulsoy and Koren (1989).

2.1.3 Purpose and Scope

This chapter reviews the research on control of machining processes conducted during the 1980s and early 1990s in order to assess the impact that this research has had on industrial practice and to discuss the expected research directions for the next decade. First the dynamic modeling of machining processes is summarized, since these models are essential for controller design and evaluation. Included are some examples from my own research work. In the next section the recent research accomplishments at the servo, process, and supervisory control levels are presented and their economic impact discussed. These developments are also illustrated with examples from my own research. Subsequently views on future research directions in automation of machining processes are given. The final section includes a brief summary and conclusions.

2.2 DYNAMIC MODELING OF MACHINING SYSTEMS

A computer-controlled machine tool is a system composed of several components: the controller, the drive servos, the machine-tool structure, the workpiece, and the cutting process. For the purpose of controller design and analysis, it is necessary to develop dynamic models of the subsystems of a computer-controlled machine tool. A brief overview of these subsystems and their dynamic models is presented in this section for the case of a CNC lathe.

2.2.1 A Computer-Controlled Machining System

A computer-controlled machine tool is illustrated schematically in Figure 2.2. This figure shows the major subsystems (i.e., the interpolator and drive servo, the machine tool and workpiece structure, and the cutting process) as well as their interactions.

Table 2.1 shows the typical magnitudes and time scales associated with major error sources from the machine tool, the control, and the cutting process. The geometric and thermal errors of the machine tool, and the forced machine vibration, dominate machining accuracy in fine cutting (i.e., finish cutting with high accuracy). During fine cutting the tool motion is very slow with small cutting forces, and structural dynamics do not affect the machining accuracy (unless the workpiece's structure is very weak).

Geometric and thermal errors, and compensations for them, have received extensive attention from researchers. Such machine tool errors can, however, be decoupled from the other error sources. For example, at the beginning of machining a new part, a cleaning cut usually eliminates any form error in the mounted workpiece. Forced machine vibrations also cannot be predicted in the early stage of machine/part design, but such effects can be reduced by balanced dynamic components and vibration isolation.

An overall machine system block diagram is given in Figure 2.2, and each subsystem is described further in the subsections below. For each subsystem we discuss the associated errors and simple dynamic models.

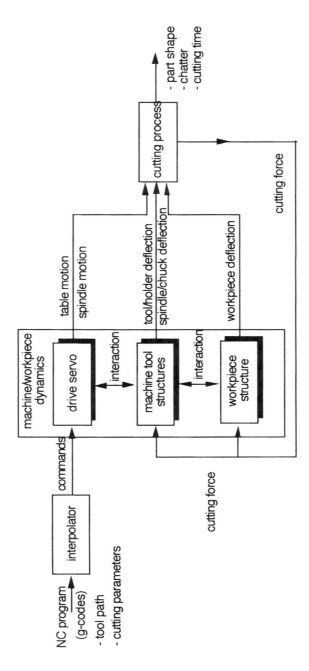

FIGURE 2.2. Configuration of a computer numerically controlled (CNC) turning system.

37

TABLE 2.1
Error Sources in Turning

Error Source	Magnitude, Maximal/Typical (μm)	Time Constant, Typical
Machine tool		
Thermal errors	250/100	∼ 1 hour
Geometric errors	100/50	∼ 1 month
Forced vibrations	200/100	∼ 10 ms
Control		
Drive servo dynamics	200/50	∼ 50 ms
Programming errors	10/5	—
Programming errors	30/10	—
Cutting process		
Chatter	400/100	∼ 10 ms
Tool wear	50/25	∼ 10 min
Machine-tool deflection	50/20	∼ 10 ms
Workpiece deflection	200/10	∼ 10 ms

Source: Adapted from Chen, Ulsoy, and Koren (1994).

2.2.2   Servo Controller and Drive System

Some sources of machining inaccuracy caused by the controller/drive dynamics are cutting force disturbances and the inertia of the drive and the machine table. The effect of these sources can be reduced by an interpolator with a deceleration function or by advanced feed drive controllers. Progamming and interpolation error sources are determined by the resolution of the CNC machine tool and therefore are small compared with other sources. The friction and backlash in the machine guideways and between the lead screw and the machine table may be serious problems for many inexpensive machine tools.

For each axis of motion of the machine tool, a position controller (e.g., PI or PID controller) typically yields the closed-loop transfer function given below. The axis position is denoted by y; u is the position command (desired position), and F is the disturbance load (typically due the cutting forces, friction, etc.);

$$Y(s) = \frac{\omega_n^2}{s^2 + 2\zeta\omega_n s + \omega_n^2} U(s) + \frac{\omega_n^2 s}{s^2 + 2\zeta\omega_n s + \omega_n^2} F(s). \qquad (1)$$

2.2.3   Machine Tool and Workpiece Structural Dynamics

Machine tool chatter is a major constraint that limits the productivity of the turning process. It is discussed in detail in other chapters of this book and illustrated schematically in Figure 2.3. The machine tool's (and/or workpiece's) structural dynamics and the cutting process interact to produce the conditions that lead to chatter.

Due to the demand for high productivity, one often selects high feed rates and deep cuts, which together induce large cutting forces. Therefore machine structural statics and dynamics are dominant factors in machining quality. Force deflections of the machine tool/toolholder and the workpiece/spindle can also significantly contribute to machining inaccuracy during heavy cutting. Such deflections can be calculated once the compliance of the machine tool is obtained.

The machine tool structure (i.e., spindle unit, tool, etc.) is typically modeled by a set of two uncoupled second-order linear equations. The relationship between tool displacements and the total forces acting on the tool may be expressed in the Laplace domain as

$$
\begin{bmatrix} x(s) \\ y(s) \end{bmatrix} = G(s) \begin{bmatrix} F_x(s) \\ F_y(s) \end{bmatrix},
\tag{2}
$$

where

$$
G(s) = \begin{bmatrix} \dfrac{(\omega_{n_x}^2/k_x)}{s^2 + 2\zeta_x \omega_{n_x} s + \omega_{n_x}^2} & 0 \\ 0 & \dfrac{(\omega_{n_y}^2 k_y)}{s^2 + 2\zeta_y \omega_{n_y} s + \omega_{n_y}^2} \end{bmatrix}.
\tag{3}
$$

The motion in the z direction (feed direction) is ignored, since the system is much stiffer in this direction than in the x and y directions and is modeled as perfectly rigid. The parameters of the transfer function matrix are determined through dynamic and static tests (Chen, Ulsoy, and Koren, 1994). In the above model development it was assumed that the machine tool is compliant and the

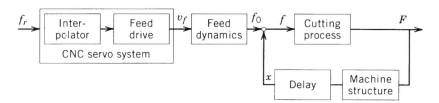

FIGURE 2.3. Block diagram of the complete machining system dynamics.

workpiece rigid. When there is a compliant structure from a slender workpiece, the workpiece deflections caused by the cutting force may become much more significant. In those cases a workpiece dynamics model can be developed as above.

2.2.4 Machining Process

The cutting process, together with the structural dynamics, contributes to the error associated with chatter as discussed in the previous section. Tool wear is important under long-term cutting conditions with high-dimensional accuracy requirements. Excessive tool wear can also contribute to tool breakage. Other error sources, like the workpiece thermal expansion due to cutting process heating, are typically small and neglected.

The machining, or metal removal, process is modeled in terms of forces being proportional to chip thickness, which in turn can be determined from the feed f, depth of cut d, and cutting speed V:

$$F \propto f^{1+\alpha} d^{1+\beta} V^{\gamma}. \tag{4}$$

As shown in Figure 2.3, the differences between the current and previous tool displacements in the x and y directions, respectively, are defined as

$$\Delta x = x(t) - x(t - T_t), \tag{5}$$

$$\Delta y = y(t) - y(t - T_t), \tag{6}$$

where T_t represents the time for one revolution. The dynamic force process, which represents changes in the forces due to changes in the tool displacements, can be approximated by the time-invariant relationship

$$\begin{bmatrix} \Delta F_x \\ \Delta F_y \end{bmatrix} = du_c A^0 \begin{bmatrix} \Delta x \\ \Delta y \end{bmatrix} \tag{7}$$

where d is the depth of cut, A^0 depends on the tool and process geometry, and for the static, nonlinear models in (4), the specific cutting energy in (7) is given by

$$u_c = K_s f^{\alpha} d^{\beta} \left(\frac{V}{1000} \right)^{\gamma} \tag{8}$$

2.2.5 Example: A Simulator for a CNC Lathe

Knowledge of interactions among the machine tool components and the cutting process is necessary to achieve improvements in machining productivity and quality. These interactions, however, have not been extensively discussed in the literature. The importance of such interactions becomes more evident with recent trends in high-speed machining and hard turning, for example. During high-speed machining, fast feed drives, rigid machine structures, high-speed spindles, and high-power spindle drives are required to maintain machining productivity. The need for fast feed drives necessitates the use of direct drive systems for the machine table and the cutting tool; it results in larger load torques that may significantly affect the system's accuracy. During hard turning, adequate machine rigidity, coupled with a rigid tool-workpiece setup, is necessary to maintain machining quality. The large cutting forces may induce tool-workpiece deflections that contribute to significant dimensional errors of the machined part. Even for conventional turning operations, such interactions are important. For example, in chatter instability where the cutting dynamics and machine structure, as well as the effects of spindle's servo dynamics, must be considered.

We consider the construction of a model that enables the simulation of the major error sources during heavy cutting, which are (1) forced deflections, (2) drive servo dynamics, and (3) machine chatter. The configuration of a CNC turning system is described in Figure 2.2. The NC part program that contains the information on the tool paths and the corresponding cutting parameters is fed into the interpolator to generate on-line reference inputs to the drive servos. The drive servos then move the cutting tool against a rotating workpiece to start the cutting. The cutting process induces cutting forces that eventually interact with the machine/workpiece dynamics. Under certain circumstances such interactions may produce significant inaccuracies and even instability of the machining process. Combining the subsystem models presented previously (after reformulation into state equation form) for the drive servo, structural dynamics, and cutting process, we can obtain a state space model of the entire CNC turning system. These equations are of the form,

$$\dot{\mathbf{x}} = \mathbf{f}_t(\mathbf{x}, \mathbf{F}_c, \mathbf{u}_c),$$
$$\mathbf{y} = \mathbf{g}_t(\mathbf{x}), \tag{9}$$

where the state vector $\mathbf{x}$ and the output vector $\mathbf{y}$ are defined as

$$\mathbf{x} = \begin{Bmatrix} \mathbf{x}_d \\ \mathbf{x}_t \\ \mathbf{x}_w \end{Bmatrix}, \quad \mathbf{y} = \begin{Bmatrix} \dot{\theta} \\ x + x_t + x_w \\ z + z_t \end{Bmatrix}. \tag{10}$$

The $\mathbf{x}_d$, $\mathbf{x}_t$, and $\mathbf{x}_w$ are the state vectors associated with the drive system, tool vibration, and workpiece vibration, respectively. The angular position of the workpiece is θ, and (x, y, z) are coordinate directions. Given a cutting force model $\mathbf{F}_c(\mathbf{y}) = \mathbf{F}_c(\mathbf{g}_t(\mathbf{x}))$, one is able to simulate the machining inaccuracy (i.e., dimensional errors of a machined part) and instability (i.e., chatter).

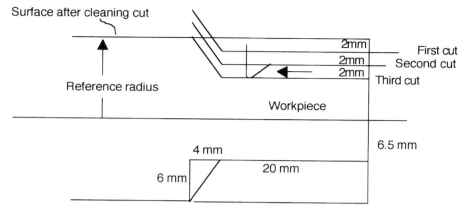

FIGURE 2.4. Experiment design for cutting force and part dimension simulations.

A CNC lathe without a tail stock was calibrated in order to assess the proposed formulation. It is driven by a 30-hp DC motor that provides spindle speeds up to 3000 rpm, and it is controlled by a CNC controller. An AISI 1020 steel bar of diameter 25.4 mm and 100 mm overhang was clamped on the CNC lathe. A Valenite 370 triangular insert was chosen such that the back rake, the side rake, the lead angle, and the nose radius are −7 deg, −7 deg, 0 deg, and 0.7938 mm, respectively. Contouring cuts were performed to shape the work-piece with a constant spindle speed of 1400 rpm, and a constant feed of 0.2 mm/rev (see Figure 2.4). The workpiece radius right after the cleaning cut was used as a reference so as to eliminate geometric errors. Then three contouring cuts were performed to shape the workpiece.

The three force components were measured and compared with the simulation results (see Figure 2.5). The machined part dimensions were also measured using a coordinate measurement machine and compared with the simulation results (Figure 2.6). These results are quite encouraging, and are discussed in detail in Chen, Ulsoy, and Koren (1994).

2.3   CONTROL OF MACHINING PROCESSES

The use of advanced control methods for manufacturing processes increased during the 1980s (Ulsoy and Koren, 1993). This can be attributed to several complementary factors: (1) the demand for better productivity, precision, and quality in manufacturing continued to increase; (2) the accomplishments of the 1970s had not solved the problems but clearly indicated the complexity of the machining system including the nonlinear, nonstationary and multivariable nature of the processes to be controlled; (3) there were significant developments in computer technology, on the one hand, and in control and estimation theory, on the other (Wright and Bourne, 1988; Ulsoy and DeVries, 1989).

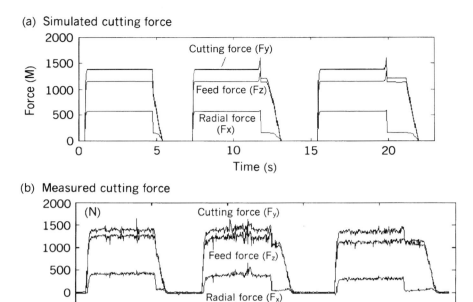

FIGURE 2.5. The simulated and measured cutting forces.

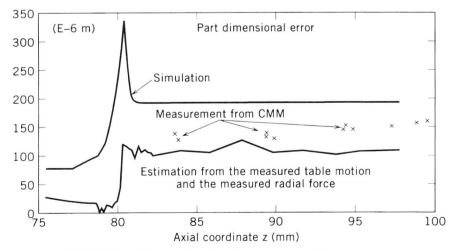

FIGURE 2.6. The simulated and measured part dimensions.

Methods such as optimal control, adaptive control, preview control, multivariable control, robust control, and state estimation have all been demonstrated
on machining processes using the inexpensive and reliable microcomputer
control technology of the 1980s. Many of these developments are applicable to
all manufacturing processes; however, our focus here will be on machining
processes such as turning, milling, drilling, and grinding.

2.3.1 Machining Control Hierarchy

Figure 2.7 schematically illustrates a convenient classification for the different
levels of control that one might encounter in a controller for a machining
process. The lowest level is servo control, where the motion of the cutting tool
relative to the workpiece is controlled. Next is a process control level where
process variables such as cutting forces and tool wear are controlled to
maintain high production rates and good part quality. Finally the highest level
is a supervisory one that directly measures product-related variables such as
part dimensions and surface roughness. The supervisory level also performs
functions such as chatter detection, tool monitoring, and machine monitoring.
The basic problems, approaches, and accomplishments at each of these control
levels will be briefly described below.

2.3.2 Servo Control

The servo level controllers can be classified into two cases: (1) point to point
(PTP), and (2) contouring. In a PTP system the controller moves the tool to
a required position where the process operation is performed. The trajectory
between the points is not controlled. Point-to-point control is required in
applications like drilling and spot welding, and is the simpler of the two cases.
It has much in common with PTP servo control problems in other application
areas, such as robotics. However, the PTP control problem in manufacturing

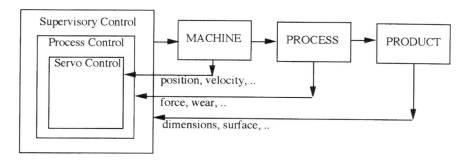

FIGURE 2.7. A controller for a manufacturing process, showing the servo, process, and
supervisory control levels.

does have some special considerations that arise from the high-accuracy requirements (e.g., 0.01 mm or better) and the interactions between the machine tool and the process as shown in Figures 2.2, 2.3, and 2.7. For example, cutting forces and heat-generated during the process must be considered to ensure accuracy in the PTP problem. The machine tool geometric errors affect PTP positioning accuracy, and are of several types: (1) volumetric errors due to slight irregularities in the components of the machine structure, (2) volumetric errors due to the machine thermal effects, and (3) volumetric errors due to the effects of cutting forces. The first type can be compensated for, based on off-line measurements (and modeling) of the machine geometric errors by technologies such as laser interferometry. However, the second and third type require on-line compensation, typically based on temperature and/or force measurements and models relating these measured variables to geometric errors in each axis of the machine tool. Typically minimum variance, forecasting, or predictive control approaches are used in these applications. Reductions in the machine geometric and thermal errors can improve accuracies by an order of magnitude (e.g., from 200 to 20 μm) in medium to large drilling and machining centers.

The contouring problem is what really makes servo level control difficult, and interesting, from a control point of view (Koren and Lo, 1992; Tomizuka, 1993). Contouring is required in processes such as milling, turning, and arc welding. The task is one of tracking trajectories (lines, parabolas, circles, etc.) very accurately in a spatial rather than temporal domain. The coordinated motion of multiple axes is required to generate accurate contours. Interpolators are used to generate the desired (reference) trajectories for each machine axis from the desired part geometry. The goal is to reduce contour errors that depend on the axial errors. The interesting problem in machining is that reducing the contour error may or may not be the same as reducing individual axis errors. Cross-coupling control (CCC) methods have been developed by Koren to couple the machine axial controllers so that contour errors rather than individual axis errors can be reduced. To decrease the tracking errors in contouring operations, feed-forward and preview controllers can also be added to the control loops. Preview controllers use the future reference trajectories that are available, since the desired part geometry is known. The concept of feedforward controllers is based on pole/zero cancellation. However, the closed-loop system may include unstable zeros which preclude the use of pole/zero cancellation methods. Tomizuka has developed a zero-phase error-tracking controller (ZPETC) that can be implemented even in the presence of unstable zeros. Control of the depth of cut, in addition to feed and speed, has also been used to produce noncircular parts on a lathe.

2.3.3 Process Control

Successful implementation at the process control and supervisory levels requires realistic process modeling based on a good physical understanding of the process to be controlled. This is potentially an area where knowledge gained from nonlinear dynamics research applied to machining processes can be useful. Some attempts at process control have utilized earlier ACO concepts (Ulsoy and Koren, 1993). For example, Watanabe reported the development of an ACO system for milling. An ACO system for grinding was proposed and demonstrated and subsequently generalized to the optimal locus control method by Koren.

The requirement for wear measurement or estimation in ACO has led to the development of model-based schemes for estimation of wear from indirect measurements such as cutting forces, temperature, and acoustic emissions. One approach to wear estimation from process measurements utilizes adaptive state estimation techniques (Ulsoy and Koren, 1989, 1993). A model of the machining process, based on the physical mechanisms of tool wear, is utilized. The tool wear is included in the model as a state variable. The measured output is the cutting force, which is used in conjunction with the model to estimate the state of wear of the tool. The difficulties in implementing this approach are due to the nonlinear nature of the process model and variation of the model parameters. Earlier attempts using a linear adaptive state observer were only partially successful; thus an adaptive nonlinear state observer is required for effective estimation of the wear. Although this wear estimation approach works well enough to be used as the basis for tool replacement strategies, even under varying cutting conditions (e.g., feed or speed), it is not sufficiently accurate for compensation of geometric errors due to tool wear. A more accurate, and flexible, tool wear monitoring system can be developed by combining the force measurement based adaptive observer with direct optical measurement of the tool wear using a computer vision system as illustrated in Figure 2.8a. The vision system is very accurate but can only be used between parts when the tool is not in contact with the workpiece. The force-based adaptive observer can be used during cutting. By intermittently (i.e., between parts) recalibrating the adaptive observer using the vision system, a very accurate estimate of tool wear can be obtained even under large variations in the cutting conditions. Experimental results, shown in Figure 2.8b, are from laboratory turning tests with large stepwise changes in the feed rate. The solid line is the tool wear estimate, the circles are the vision measurements used to recalibrate the observer, and the x's are the wear values measured using a toolmaker's microscope. Errors in the flank wear estimate are less than $50\ \mu m$ (2×10^{-3} inches) over the entire cutting period, and they improve as the cutting proceeds toward the end of the tool life.

Suboptimal process control systems were considered to be more practical than ACO systems, and systems of the ACC or GAC type were developed for various applications (Ulsoy, Koren, and Rasmussen, 1983). The emphasis has been on ACC systems, typically based on force or torque measurements, rather than GAC systems, which are based on product dimensions and surface finish. This is primarily a result of the availability of less expensive and more reliable sensors for force and torque measurement. However, with recent advances in

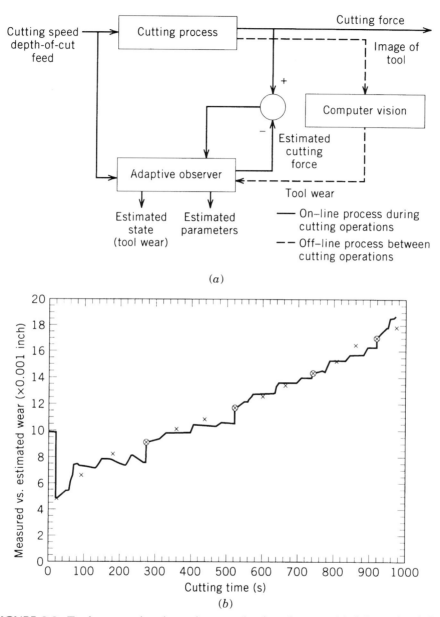

FIGURE 2.8. Tool wear estimation using an adaptive observer: (*a*) Schematic of the adaptive observer integrated with computer vision. (*b*) Estimated and measured flank wear in a turning experiment under varying cutting conditions (× microscope measurement, ⊗ computer vision calibration of the observer, —— estimated flank wear from the adaptive observer).

optical sensing techniques for part dimensons and surface finish, this trend can be expected to change in the coming decade. Use of optical sensing in GAC systems has already been demonstrated in turning and milling applications (Ulsoy and Koren, 1993).

Despite developments in ACC systems, machining processes were still found to be difficult to control because of process parameter variations, which can lead to problems such as tool breakage or even instability. Consequently adaptive control methods were widely investigated for a variety of machining processes (Ulsoy, Koren, and Rasmussen, 1983). The basic problem arises from the fact that the effective process gains and time constants depend on process variables such as feed, speed, and depth of cut. Furthermore process-related parameters such as tool geometry, work, and tool material properties are difficult to characterize and can have a significant effect on the process dynamics. This unpredictability of the process can lead to poor performance, tool breakage, or even instability when fixed gain process controllers are used. These problems, due to process model uncertainty, were a primary reason for the poor industrial acceptance of the first generation of AC systems for machining. To address this problem, adaptive control schemes, combining on-line parameter estimation and control, were developed and applied to machining processes. As an example of this class of controllers, a model reference adaptive force controller for a milling process is illustrated in Figure 2.9*a* and described below.

A discrete-time model of a two-axis slot-milling process can be written as (Ulsoy and Koren, 1989, 1993)

$$\frac{F(z)}{V(z)} = \frac{b_0 z + b_1}{z^2 + a_1 z + a_2},$$ (11)

where $V(k)$ is a voltage signal proportional to the machine feedrate, and $F(k)$ is the measured resultant force. The parameters a_1, a_2, b_0, and b_1 depend on the spindle speed, feed, depth of cut, workpiece material properties, and so on. Thus they must be estimated on-line for effective control of the resultant force based on manipulation of the feedrate (i.e., $V(k)$). The goal is to maintain the resultant force at the reference level, $R(k)$, which is selected to maintain high metal removal rates without problems of tool breakage. The process zero at $-(b_1/b_0)$, although the values of b_0 and b_1 can vary, was found to be inside the unit circle for all cutting situations at the 50-ms sampling period that was used. Therefore direct adaptive control methods can be employed. A model reference adaptive controller was designed and evaluated in laboratory tests for two different workpiece materials and for changing depths of cut. The controller design was straightforward and gave satisfactory performance despite the process variations. The main difficulty is eliminating bias in the parameter estimates due to the "runout" noise in the force measurements. Experimental results are shown in Figure 2.9*b* for machining of a 1020 CR steel workpiece with step changes in depth of cut from 2, 3, to 4 mm. The desired value of

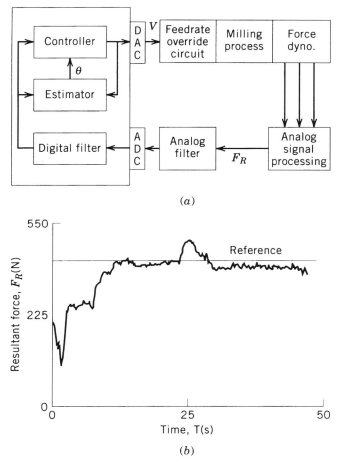

(a)

(b)

FIGURE 2.9. Model reference adaptive force control in milling: (a) Schematic of the experimental system. (b) Resultant force versus time with MRAC in slot milling of a steel workpiece with step changes in depth of cut.

$R = 400 \text{ N}$ is maintained despite these large changes in depth of cut at a spindle speed of 550 rev/min. Note that initially the feed rate saturates, and the reference value of 400 N cannot be achieved at the 2-mm depth of cut. Also initial transients occur due to the parameter adaptation. This controller also gave good results when used in machining other materials (e.g., aluminum) without any additional controller turning.

2.3.4 Supervisory Control

The supervisory level is the area that has received the least research attention to date and can be expected to be a major focus of future research. There are in fact many levels of control in a machining process control hierarchy, and what we have referred to here as supervisory control includes functions such as selection of control strategies, sensor fusion, generation of reference commands for the process control level, tool breakage monitoring, chatter detection, and machine monitoring. Supervisory level strategies are aimed at compensating for factors not explicitly considered in the design of the servo and process level controllers. For example, knowledge of the part geometry as contained in the CAD system can be used to determine the reference values of process variables. Furthermore the supervisory level can monitor for tool and/or machine failures or degradation. Process-related phenomena, such as chatter and chip entanglement, can also be monitored. Information from various process-related sensors can be integrated to improve the reliability and quality of sensor information. Finally all of this information can be used to achieve on-line optimization of the machining process. This is an area where knowledge gained from nonlinear dynamics can be potentially useful for control, particularly with respect to chatter prediction.

As an example of supervisory control in machining, consider a through-hole drilling operation. A successful control strategy must ensure good hole quality, prevent drill breakage, minimize burr formation at the exit, and also maintain high metal removal rates. A strategy that achieves these goals includes a servo control level where the spindle speed and the drill feed rate are accurately controlled. A process control level implements a torque controller to prevent drill breakage. The supervisory controller, based on the drill position (depth of hole), selects an appropriate strategy for use at the process control level as well as the appropriate reference signals for that strategy. Specifically the supervisory level selects a feed and speed control strategy at the entrance phase based on hole quality (hole location error) considerations. The feed and speed references are selected based on the required hole location error constraint. After the hole is initiated, the supervisory controller switches to a speed and torque control strategy with the reference values determined from tool breakage and tool wear considerations. Finally, as the operation nears drill breakthrough, the supervisory controller switches back to a feed and speed control strategy to minimize burr formation. Thus the supervisory controller selects control strategies and their associated reference levels for optimization of the drilling process.

The results of using such a supervisory control strategy are illustrated in Table 2.2. This table compares experimental results obtained using four types of controllers: (1) no control (conventional approach where nominal feed and speed values are selected but not controlled during drilling), (2) feedback control of feed and speed, (3) feedback control of torque and speed, and (4) the supervisory controller (which combines 2 and 3 as described above). The comparison is made in terms of (1) machining time for one hole, (2) burr rating, (3) hole location error (in terms of the pooled standard deviation of the hole location error), and (4) percentage of holes drilled with stoppage events. The machining time per hole is given in seconds, the burr ratings ranges from 1

TABLE 2.2
Control Strategy Comparison in Drilling

	No Controller	Feed/Speed Controller	Torque/Speed Controller	Supervisory Controller
Machining time (s)	11.11	11.28	9.79	11.71
Burr rating	2.93	2.94	2.26	1.58
Pooled standard deviation ($\times 10^{-3}$ inches)	4.43	4.53	6.28	4.25
Stoppage events (%)	25	15	0	0

Source: Adapted from Ulsoy and Koren (1993).

(very little burr formation) to 5 (large burrs), and the hole quality is given in terms of a pooled standard deviation with smaller values indicating better hole location accuracy. The percentage of stoppage events is for holes for which the torque has exceeded a maximum allowable value for drill breakage and the process was stopped. In this application a hole location error pooled standard deviation of $< 4.5 \times 10^{-3}$ inches and a burr rating of < 1.75 are required. It is also desired that there be no stoppage events and that the machining time be minimized subject to these hole quality, burr, and breakage constraints. The supervisory control strategy is the only one that meets the required hole quality and burr constraints, while eliminating stoppage events. It also yields a machining time comparable to the uncontrolled or feed/speed control cases and only slightly longer than the torque/speed controller. The results are average values based on a statistical study involving the drilling of 20 holes with each strategy in a randomized order. These results clearly illustrate the potential advantages of a supervisory control strategy over each of the individual process control strategies (i.e., feed, speed, or torque control).

2.4 INDUSTRIAL PRACTICE AND FUTURE DIRECTIONS

The research advances of the 1970s and 1980s have not had a serious impact to date on industrial practice, despite significant findings and numerous laboratory demonstrations. This is in part due to the unfavorable experience with the first generation of AC systems and their components introduced in the 1970s, and in part because these systems are still difficult to apply without considerable knowledge and experience on the part of the operator. One of the trends we can expect to see in the 1990s is the industrial application of many of the ideas demonstrated in laboratories in the 1970s and 1980s. The problems with process parameter variations have been addressed by the research of the 1980s as described previously. The technology, however, is still difficult to apply, and this is an area of on-going research. Global competition will force the machine tool industry, which in the United States took a serious economic downturn in the 1980s, to develop new state-of-the-art products. These new products will, in general, be machines whose accuracy and productivity is enhanced through the use of sensing and control technology. Thus machine tool builders will need to upgrade the capabilities of their products by adding sensors, controls, and software.

2.4.1 Implementation Issues

A key impediment to the industrial use of research advances has been the problems associated with system integration, upgradability of systems, and ease of use (e.g., complexity of programming). The various advances in control of machining processes, although demonstrated in laboratory systems, have been difficult to integrate into commercial systems. Controllers for CNC machine tools are closed systems where both software and hardware are specifically engineered for each system. As a result the components produced by different vendors cannot be integrated into a system for a particular machine tool application. These systems are difficult to upgrade as new technology becomes available, and new technology (e.g., new AC systems for machining) cannot be readily integrated into existing commercial systems. In response to this problem, the next generation controller (NGC) project was initiated. It attempts to develop a standard open architecture for machine tool controllers. The acceptance of this standard will be an important step toward improved control of machining processes.

In-process sensing and control techniques being developed today will become the key component of the next generation of quality control (see Figure 2.10). Currently quality is ensured in the product engineering cycle at two stages. The first uses Taguchi-type methods at the product design stage to ensure that quality is designed into the product. Obviously this is done *before* the part is manufactured. The second uses statistical process control (SPC) methods at the inspection stage, *after* the part is manufactured, to check the quality of the manufactured part. However, real-time sensing and control will introduce a third level of quality assurance that can be implemented *during*

FIGURE 2.10. The next generation of quality control will involve quality assurance not only at the design and inspection stages but also in-process quality control implemented at the machining stage.

machining (i.e., in-process). This will complement Taguchi and SPC methods, lead to the next generation of quality control, and eliminate the need for expensive postprocess inspection. Such in-process quality assurance methods are currently being employed in many processes. However, widespread implementation in machining has not been achieved due to the required trade-offs in machining among quality, productivity, and cost.

The hierarchical control structure illustrated in Figure 2.7 considerably simplifies machining processes, since they have many (not three) levels of control, each with its own characteristic interation time. For example, servo control (axis motion control) loops may have sampling periods of 3 ms, process (e.g., force or torque) control loops may have sampling times of 30 ms, control of wear rate may require sampling at 3 s, tool changes may occur at intervals of 30 min, predictive machine maintenance at intervals of 30 days, and so on. Typically, for predictable and efficient operation, all of these levels must be considered in designing a controller for a machining process. For example, performing predictive machine maintenance will improve the performance of the servo loops, and changing worn tools will improve the process controller performance. A multilevel hierarchical control scheme will make the process more predictable at the lower control levels, thus reducing the controller complexity and control effort required at those lower levels. The design of such multilevel controllers will require a good understanding of all aspects of the machining process in order to include them in appropriate models. Future research on control of machining processes can be expected to be aimed at these higher levels in the control hierarchy.

2.4.2 Future Directions—Artificial Intelligence

Methodologies developed in artificial intelligence, expert systems, fuzzy control, and neural networks are expected to be useful at the higher levels in the control hierarchy for machining processes, and to complement traditional methods from control and estimation theory. Fast optical sensing methods will allow us to close the loop on part dimensions, surface finish, and wear, and not only process variables like force or tool holder vibrations. These developments in sensing will be the necessary ingredient for in-process quality control (see Figure 2.10). Along as the developments in computer software and hardware toward modularity and an open-system architecture will come software modules for rapid modeling and compensation for various error sources (e.g., friction, backlash, thermal deformations, geometric errors). The trend in machining is toward higher speeds, and this trend will require appropriate control strategies associated with machining processes (e.g., cutting will occur during transients in contouring operations). One can expect to see the fruits of current efforts toward standardization in hardware, software, and interfacing in the 1990s. These technological developments will be extremely important for the future of the machine tool industry and its economic viability.

2.4.3   Future Directions—Nonlinear Dynamics and Chaos

Another rapidly developing field is nonlinear dynamics, and its related topics such as chaos, fractals, complex systems, nonlinear signal processing, and control of chaotic systems (Moon, 1992). Although much interest has been generated, so far these developments have had little impact on control of machining processes (Kumar, 1995). Other chapters in this book discuss some more promising application areas, such as machining chatter and surface characterization. Here I will comment briefly on control of chaotic systems (Shinbrot *et al.*, 1993) and their potential utility in machining process control.

The basic idea of control of chaotic systems is based on the work of Ott, Grebogi, and Yorke and has come to be called the OGY method. For a chaotic system, no matter how small a finite control we are willing to make, the orbit will eventually come close enough to a fixed point that our control will bring it to the fixed point. Once at the fixed point, it only takes minimal control action to keep it there. Furthermore, since a chaotic system has an infinite number of such fixed points, the OGY control (with very little control effort) can easily stabilize the system at different fixed points. Such an approach is not likely to be of direct practical value in machining systems, since we do not normally want to operate in a chaotic state (e.g., chatter). A preliminary study combining chaos theory and neural network controllers for chatter control has been reported (Kumar, 1995).

2.5   SUMMARY AND CONCLUSIONS

This chapter has summarized the important research results of the past decade in the dynamic modeling and control of machining processes. Although the machine controller industry is relatively small, control of machining processes can have a significant economic impact. Sensing and control technologies have played an increasingly important role in machining since the introduction of CNC systems in the 1960s. At the servo control level advances in both point-to-point and contouring control cases have been discussed. These include the introduction of feed-forward and preview techniques as well as the cross-coupling controller which minimizes contour errors rather than individual axis errors. At the process control level, where process variables such as force or torque are controlled, adaptive techniques have been employed to address the problem of process parameter variations and uncertainty. At the supervisory control level additional control and monitoring capabilities (e.g., sensor fusion, chatter detection, tool and machine condition monitoring, on-line process optimization) have been incorporated.

The impact of these research developments on industrial practice, most of them demonstrated in the laboratory, has been modest. However, the economic factors driving the research on control of machining are very strong. Thus widespread industrial implementation of many of the research developments reported here can be anticipated in the coming decade. A key to enabling widespread industrial application of sensing and control technologies in machining is the development of open-architecture systems that facilitate the integration of new technologies into commercial products in a cost-effective manner. Such systems hold the promise of achieving in-process quality assurance, system flexibility, and high production rates in a cost-effective manner.

REFERENCES

Chen, S. G., Ulsoy, A. G., and Koren, Y. 1994. Machining error source diagnostics using a turning process simulator. *Proc. Jap.–USA Symp. on Flexible Automation*, Kobe, Japan, July 1994, pp. 275–282.

Koren, Y. 1983. *Computer Control of Manufacturing Systems*. McGraw-Hill, New York.

Koren, Y., and Lo, C. C. 1992. Advanced controllers for feed drives. Keynote paper presented at the CIRP Assembly. *CIRP Ann.* **2**.

Koren, Y., and Ulsoy, A. G. 1989. Adaptive control in machining. In Davis J. R., ed., *Metals Handbook: Machining*, vol. 16, 9th ed. ASM International, Metals Park, OH, pp. 618–626.

Kumar, S. R. T. 1995. Monitoring and control of manufacturing processes using chaos theory. Technical report. Intelligent Design and Diagnostics Research Laboratory, Pennsylvania State University, University Park.

Moon, F. C. 1992. *Chaotic and Fractal Dynamics*. Wiley, New York.

Shinbrot, T., Grebogi, C., Ott, E., and Yorke, J. 1993. Using small perturbations to control chaos. *Nature* **363**:411–417.

Tomizuka, M. 1993. On the design of digital tracking controllers. *ASME J. Dynamic Sys., Meas. Control*, Special 50th Anniversary Issue, June 1993.

Ulsoy, A. G., and Koren, Y. 1989. Applications of adaptive control to machine tool process control. *IEEE Control Sys.* **9**:33–37.

Ulsoy, A. G., and Koren, Y. 1993. Control of machining processes. *ASME J. Dynamic Sys., Meas. Control*, Special 50th Anniversary Issue, June 1993.

Ulsoy, A. G., and DeVries, W. R. 1989. *Microcomputer Applications in Manufacturing*. Wiley, New York.

Ulsoy, A. G., Koren, Y., and Rasmussen, F. 1983. Principal developments in the adaptive control of machine tools. *ASME J. Dynamic Sys., Meas. Control* **105**:107–112.

Wright, P. K., and Bourne, D. A. 1988. *Manufacturing Intelligence*. Addison-Wesley, Reading, MA.

3

DYNAMIC PROBLEMS
IN HARD-TURNING,
MILLING, AND
GRINDING

M. A. DAVIES

3.1 INTRODUCTION

In this chapter we examine the application of nonlinear dynamics to various
problems in manufacturing. We focus on three practical machining processes
currently receiving significant industrial attention: turning of hardened steels
(hard-turning), high-speed milling, and high-speed grinding. In each of these
areas, researchers are seeking to expand the envelope of machinable materials,
reduce manufacturing costs by increasing material removal rates, and manufacture parts with more complex shapes, while maintaining or increasing
current levels of accuracy. Attaining these goals requires manufacturers to
operate at the limits of their machine tools, pushing spindle speeds and feed
rates as high as possible to increase machining speeds and lower manufacturing
costs. Unlike conventional machining operations, which could reasonably be
considered to be quasi-static, these modern manufacturing processes are highly
dynamic. In addition strong nonlinearities can arise from the cutting process
and the behavior of the machine-tool, and these nonlinearities can have
unpredictable effects on the process dynamics. If these effects could be accurately modeled, significant improvement in performance might be attained by
including control and compensation algorithms in machine-tool controllers.

Dynamics and Chaos in Manufacturing Processes, Edited by Francis C. Moon.
ISBN 0-471-15293-5 © 1998 John Wiley & Sons, Inc.

This chapter will discuss each manufacturing area noted above, and point out through examples where the techniques of modern nonlinear dynamics may be applied to improve the understanding, control, and performance of the process.

3.2   HARD-TURNING

Many precision components that must resist wear in harsh mechanical and thermal environments are manufactured from heat-treated hardened steels with Rockwell-C hardness ratings ranging from 50 to 65. Examples include gears, bearing components, and molds for plastic lenses. The global market for machining hardened steel parts is estimated to be in excess of $150 million dollars per year (Mason, 1992; Noaker, 1992), but this market might be expected to grow substantially if the process were to become more reliable. Until recently grinding was the only economical alternative for machining hardened steels with the required tolerances. However, the development of more wear-resistant tool materials such as polycrystalline cubic boron nitride (PCBN) and ceramics have made other forms of machining possible. Currently manufacturers are successfully turning, boring, milling, and broaching heat-treated hardened steel components (Konig et al., 1990). Replacing more costly and less agile grinding operations (Konig et al., 1993) with alternative machining operations has the potential to substantially reduce production costs.

The term "hard-cutting" (of which "hard-turning" is a subset) refers to the machining of various types of hardened ferrous metals using tools with "geometrically defined cutting edges" (Konig et al., 1990). (Grinding does not fall in this category, since the number and orientation of the abrasive particles that do the cutting are unknown.) Metals that can be machined in this manner include AISI 52100 bearing steels, AISI M50 die and tool steels, nickel-iron superalloys, and some hard cast irons. In terms of wear rates and surface quality, PCBN and ceramic tools perform well in hard-turning. PCBN tools consist of cubic boron nitride with either a metallic or ceramic binder. They are available with variable CBN content and grain size. Typical ceramic tools consist of titanium carbide (TiC) in an alumina (Al_2O_3) matrix. Although they are more expensive, PCBN tools can perform significantly better than ceramic tools in hard-turning (Konig et al., 1990). High CBN (approximately 90% by volume) tools have been found to wear more slowly in rough machining, while low CBN (70% by volume or less) tools perform better in finish machining operations (Chou and Evans, 1996). Chou and Evans (1996) have shown that the microstructure of the workpiece and CBN grain size can also have a substantial effect on tool life. Typical cutting speeds used in production range from 2 to 4 m/s, with the depth of cut ranging from 0.5 mm in roughing operations to tens of micrometers or less for finishing cuts.

Despite significant progress in transferring hard-turning operations out of the laboratory and onto the shop floor, challenges still remain (Soons and Yaniv, 1966). Two of these challenges, increasing the vibrational stability of the process and prediction of the cutting forces, are nonlinear and dynamic in nature. The primary sources of vibrational instabilities in hard-turning are regenerative chatter and stick-slip oscillations excited by the rubbing of the tool on the surface of the workpiece as it wears. Nonlinearities can arise from several sources, including the geometry of the cutting tool, friction, the cutting mechanics, and the mechanical behavior of the machine tool. Because of these nonlinearities both periodic and aperiodic vibrations are possible. For most machines used in hard-turning, the frequency range of these oscillations is a few hundred to a few thousand hertz, and is usually related to one of the lower linear modes of the machine-tool system. An entirely different type of nonlinear oscillation arises from the formation of the chip during hard-turning. Hardened steels, like many other difficult-to-machine materials (Komanduri et al., 1982; Komanduri and Brown, 1986, 1981) form segmented chips at relatively low cutting speeds. Research at the National Institute of Standards and Technology (NIST) indicates that chip segmentation leads to both periodic and aperiodic oscillations in the cutting forces and that the dynamic nature of these oscillations depends on the cutting parameters. For typical hard-turning operations, the frequency of these oscillations can be extremely high, often exceeding one hundred kilohertz. The mechanisms of chip formation in hardened steels are still not fully understood, and it seems likely that a full nonlinear dynamic model of the thermoplastic flow of the material will be necessary if cutting forces, surface finish and surface integrity are to be accurately predicted. These dynamic problems will be discussed in more detail below. Since many of the other chapters in this book discuss regenerative instabilities, in this section the emphasis will be placed on the dynamics of chip formation.

3.2.1   Regenerative Chatter in Hard-Turning

Figure 3.1 shows the specific cutting energy for orthogonal turning of AISI 52100 bearing steel using a low CBN content tool[1] for various cutting geometries. The range of specific energies is approximately 6 to 11 kN/mm^2 (Davies et al., 1996a); by comparison, more conventional steels have specific cutting energies ranging from 2.7 to 9.3 kN/mm^2 (Kalpakjian, 1991). The high specific cutting energies of hardened steels lower the critical depth (or width) of cut for which regenerative chatter occurs. Thus, to maintain stability at reasonable metal removal rates, the stiffness and damping of the machine-tool system must be high. The need to maintain high stiffness increases the cost of machine-tools and severely limits the geometry of the parts that can be machined. If the stability of the system could be increased, either by passive or active control of the process dynamics, the application of hard-turning processes could be greatly expanded. There have been numerous attempts to

---

[1]The tool consisted of 70% CBN by volume in a titanium nitride binder. The average grain size of the CBN was approximately 0.5 μm.

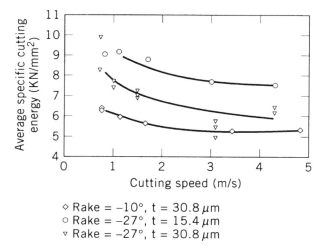

◇ Rake = −10°, t = 30.8 μm
○ Rake = −27°, t = 15.4 μm
▽ Rake = −27°, t = 30.8 μm

FIGURE 3.1. Plot of the specific cutting energy versus speed for orthogonal turning of 52100 bearing steel for various cutting geometries.

control chatter in turning and boring operations. Nachtigal et al. (1970) were among the first researchers to apply active control techniques to improve stability in turning. In their work they use an electrohydraulic actuator and cutting force feedback to improve the stability of an engine lathe. Klein et al. (1975a, 1975b) developed the theory for the control of chatter in a boring bar, and these techniques have been successfully employed in an experimental system. More recent work by Shirashi et al. (1991) and Tewani et al. (1995) discuss the application of modern control techniques in chatter suppression. Abler et al. (1995) have applied some of these techniques to the control of chatter in turning hard materials on a vertical lathe using piezoelectric actuators. This work demonstrates the significant impact that vibration control can have on system performance. However, to this author's knowledge, there have been no attempts to apply these techniques in finish hard-turning. In hard-turning operations the small nose radius of the tool and the thermoplastic behavior of the material during cutting will cause significant nonlinearities. Also rapid tool wear will cause the dynamic behavior of the system to vary in time. Any successful control system will need to address these sources of nonstationarity and nonlinearity in the process.

Detection of tool wear is another important issue in hard-turning. It is a common observation that the amplitude of tool vibration increases as the tool wears. By monitoring these vibrations, numerous researchers have attempted to predict imminent failure of the tool. However, many of these studies simply apply signal processing techniques to measured vibrations with little attention paid to the physical cause of these vibrations. Insight into the physical causes

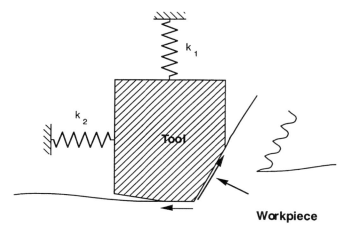

FIGURE 3.2. Schematic of tool with significant flank wear that causes the nose to rub on the workpiece.

of wear-induced vibrations could lead to more efficient monitoring of the tool condition. For example, one possible cause of increased vibrations is friction between the worn tool nose and the moving workpiece. Figure 3.2 shows a schematic of a tool that has experienced significant wear along the relief face (i.e., flank wear). Because of the wear the tool nose will rub on the workpiece and experience a drag or friction force which could lead to self-excited oscillations of the tool. Feeny (1989) has shown that nonlinearities arising in stick-slip friction can lead to complex dynamic behavior, and this work might be a useful starting point for characterizing the behavior of a worn tool. It may be possible to predict changes in the dynamic behavior resulting from increases in the contact area between the tool and workpiece with relatively simple models. These models might then be used to determine the most effective signal-processing techniques for monitoring wear. These may include nontraditional measures such as fractal dimension and Lyapunov exponents.

3.2.2 Dynamical Issues in Chip Formation

Chip formation is a highly dynamic and nonlinear process involving a very complex thermoplastic flow of material. In general, these dynamics must be understood before accurate predictions of important processes parameters such as cutting forces and temperatures can be made. Material flow patterns in hard-turning become highly inhomogeneous at low cutting speeds. This inhomogeneity manifests itself in segmented chip formation. There is some debate about the cause of this segmentation in hardened steels. Shaw (1993) and Konig et al. (1993) suggest that chip segmentation is due to periodic fracture of the workpiece material while Davies et al. (1995, 1996a) show evidence of strain localization due to thermoplastic instabilities in the material flow. Although fracture can play an important role in chip formation, research at NIST has shown (Davies et al., 1996a) that in most finish hard-turning operations, strain localization is a more prominent effect.

Strain localization has various physical causes (Anand, 1984; Asaro and Needleman, 1984). At high strain rates, such as those encountered in machining, the dominant cause of strain localization is an instability in the thermoplastic behavior of the material. The instability develops from inhomogeneities in the thermal field that grow due to the lack of time for thermal diffusion (Clifton, 1984). Thermal softening weakens the material in the regions of higher temperature, and this causes variations in the strain-rate field. In softer regions, where rapid shearing occurs, local temperatures rise due to the transformation of mechanical work into heat. This tends to increase strain rates in the regions of higher temperature and decrease strain rates in the regions of lower temperature, leading to unstable growth of the inhomogeneities. Increases in the rates of strain hardening and thermal diffusion tend to oppose the formation of instabilities because of their tendency to smooth out the inhomogeneities. Other effects such as strain-rate sensitivity can also affect the growth of the instabilities.

A large body of literature exists on the phenomenon of strain localization at high rates of strain. Zener and Holloman (1944) were the first to recognize the destabilizing effect of thermal softening during high-strain-rate deformations. Since then, most experimental studies such as those by Costin et al. (1979), Hartley et al. (1987), and Marchand and Duffy (1988) have focused on measurements of strain localization in high-strain-rate torsion (Kolsky bar) tests. In parallel there has been substantial interest in both the analytical (Clifton et al., 1984; Bai, 1982; Burns, 1985; Wright and Walter, 1987) and numerical (Litonski, 1977; Wright and Walter, 1987; Molinari and Clifton, 1987) prediction of the onset and growth of shear instabilities.

Recht (1964) is generally credited with first suggesting that thermoplastic shear instabilities are responsible for the formation of segmented chips in machining. Recht developed a model for the simple shear of a one-dimensional slab of material and predicted the critical strain rate at which instabilities will occur. This simplified model is used to predict the critical cutting speed for the formation of segmented chips in machining of mild steel.[2] Komanduri et al. (1981) and Komanduri and Brown (1982, 1986) have experimentally demonstrated thermoplastic strain localization in machining titanium alloys, AISI4340 steel, and nickel-iron alloys. They showed that polished and etched chip cross sections clearly have highly localized deformations that closely resemble shear flow in a fluid boundary layer. Marusuch and Ortiz (1995a, 1995b) have developed finite-element simulations that show the formation of segmented chips in high-speed machining of mild steel. These simulations accurately predict steady state cutting force data obtained from experiments.

---

[2]Recht uses an initial shear yield strength of 140 MPa for mild steel. By contrast, a hardened bearing steel like that discussed here would have a shear yield strength of up to six times larger.

Despite these experimental and numerical successes, there have been few attempts (Zhen-Bin and Komanduri, 1995; Semiatin and Rao, 1983) to analytically examine the formation of segmented chips. Such an analysis is difficult because, as suggested by Davies et al. (1996a), it requires consideration of the dynamic interaction between shear bands. This subject has received little attention. Recent work by Burns (1994), demonstrating the existence of Hopf bifurcation in the dynamic equations describing deformation of a bar during a low-temperature tensile test, and new analytical work by Olmstead et al. (1994) and Glimm et al. (1993), in which shear bands are treated as planar discontinuities in the velocity field, may be helpful in developing an analytical model for chip formation. Such a model might provide insights into the dynamics of segmented chip formation that cannot be deduced from experiments and complicated numerical simulations.

The remainder of this section focuses on the formation of segmented chips in hard-turning. Experimental evidence is presented that suggests that the sequential formation of localized shear zones in chips undergoes a transition from aperiodic to highly periodic behavior as the speed is increased. Although the physical causes of this transition are not currently understood, the following two hypothesis are proposed:

- The transition is due to a change in the mechanism of localizations as the nominal strain rate is increased. For example, Ramalingam et al. (1973) and von Turkovich (1971) developed a theory based on dislocation generation and annihilation to explain the fine scale localization observed in all machining chips. In hard-turning this mechanism may be responsible for nonperiodic segmentation at low speeds, while catastrophic shear localization may cause the periodic formation of segments at higher speeds.
- A nonlinear dynamic system governs the sequential formation catastrophic shear zones, and the transition in behavior is the result of bifurcations in the behavior of this system.

In the opinion of the author, the second hypothesis has the most potential to explain observed effects with the simplest physical models.

Orthogonal Cutting and Chip Mechanics

To better understand chip morphology in hard-turning, NIST conducted a set of cutting experiments on a 52100 bearing steel which was through-hardened to approximately 62 Rockwell-C. Experiments were conducted on a two-axis diamond turning machine with an air-bearing spindle. A simplified orthogonal cutting geometry was obtained by machining the edge of a 0.5-mm-wide, 94-mm diameter ring using custom ground and lapped, flat-nosed PCBN tools (see Figure 3.3a). The rake angle α, uncut chip thickness t_c, and cutting speed are defined as shown in Figure 3.3b. Experiments were conducted with the

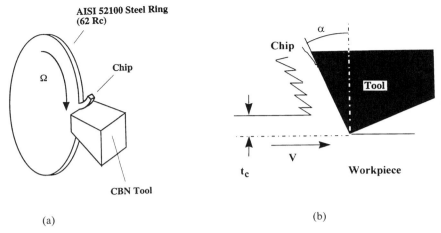

(a)   (b)

FIGURE 3.3. (*a*) Schematic of orthogonal cutting operation where a flat tool is used to remove a chip from a thin disk. (*b*) Definition of the cutting parameters in orthogonal cutting experiments.

conditions defined in Table 3.1. The tool holder was mounted on a three-axis dynamometer, and forces parallel (F_c) and perpendicular (F_t) to the cutting direction were measured. Chips were mounted, polished in cross section, and etched using a 2% Nital[3] solution.

Experimental Results

Examination of segmentation patterns in chips obtained from orthogonal cutting experiments described above showed evidence of a transition from more closely spaced but disordered patterns at lower cutting speeds to more widely spaced but periodic patterns at higher cutting speeds. Figure 3.4 illustrates this transition in SEM photomicrographs of four chips obtained with identical cutting geometry (cutting conditions 1 in Table 3.1) but with cutting speeds ranging from 3.1 m/s in panel *a* to 0.35 m/s in panel *d*. The segment spacing increases with cutting speed and is much less regular at the lower cutting speeds. Segment spacing was also affected by other factors such as rake angle and depth of cut. These effects are summarized by Figure 3.5, showing plots of mean segment spacing versus speed for two different rake angles and two uncut chip thicknesses.

Information about material flow patterns occurring during machining can be obtained from polished and etched chip cross sections. Figure 3.6*a* shows an etched cross section of a highly segmented chip formed at $V_c = 1.0$ m/s, $t_c = 31\,\mu$m, and $\alpha = -27°$. Figure 3.6*b* shows a magnified view of the region between two segments. Highly localized shear flow is evident. It is hypothesized

---

[3]Nital is a standard etchant consisting of 2% nitric acid in ethyl alcohol.

TABLE 3.1 _____
Combinations of Cutting Conditions Used in the Orthogonal Cutting Experiments

Label	Rake Angle	Uncut Chip Thickness (μm)	Cutting Speed Range (m/s)
1	$-27°$	31	0.35–3.1
2	$-27°$	15	0.35–3.1
3	$-10°$	31	0.35–3.1

that the segmented chips form as a result of catastrophic shear failure (Recht, 1964) rather than fracture (Shaw, 1993; Konig et al., 1990). Wright and Walter (1987) refer to this type of flow pattern as a central boundary layer and treat its formation with asymptotic methods. Since it is a strain-rate-sensitive process, the onset of segmented chips occurs when the cutting speed exceeds a critical value determined by material properties. Thus, for a given cutting geometry, the onset of shear localization and the resulting segmented chips are typically controlled by the cutting speed, as can be observed in Figure 3.5.

Insight into the dynamics of chip formation can be gained by considering the chip as a spatial record of the temporal dynamics during cutting, digitizing images of these chips and analyzing them with signal-processing techniques. Figure 3.7a shows profiles of serrated chip edges obtained by digitizing chip cross sections. The lower signal is from a chip obtained at $V_c = 3.1$ m/s,

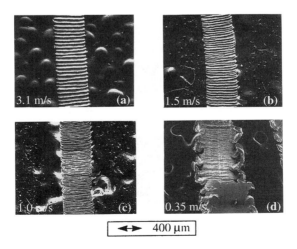

FIGURE 3.4. Scanning electron micrographs of chips obtained from orthogonal cutting with an uncut chip thickness of 31 μm, a rake angle of $-27°$ and cutting speeds of (a) 3.1 m/s, (b) 1.5 m/s, (c) 1.0 m/s, and (d) 0.35 m/s.

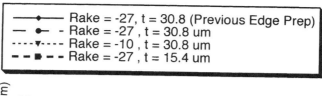

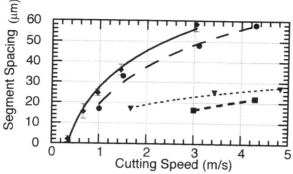

FIGURE 3.5. Mean segment spacing plotted as a function of cutting speed for various cutting geometries.

$\alpha = -27°$, and $t_c = 31\ \mu m$. The upper trace is from a chip cut with depth of cut and rake angle but with a cutting speed $V_c = 1.0\ m/s$. The power spectra of the two chips (Figure 3.7b) indicate the periodic nature of the lower chip trace and the aperiodic nature of the upper trace. The trend from ordered to disordered segment spacing with decreasing cutting speeds was observed for all cutting geometries. Similar trends are seen in the literature (Komanduri and

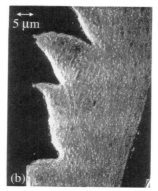

FIGURE 3.6. Low-magnification (a) and high-magnification (b) scanning electron micrographs of a chip cross section etched in 2% Nital solution. Cutting conditions were orthogonal with $V_c = 1.0\ m/s$, $\alpha = -27°$, and $t = 31\ \mu m$.

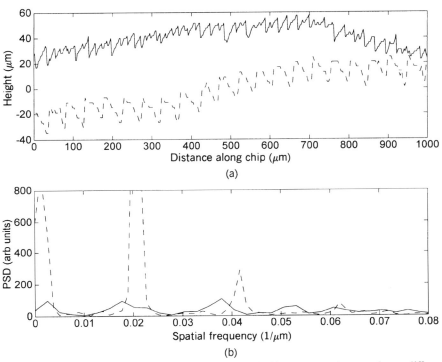

FIGURE 3.7. (*a*) Chip profiles taken from scanned chip cross sections and two different speeds. The upper trace is from a chip formed with cutting speed 1.0 m/s, while the lower trace is from a chip cut at 3.1 m/s. (*b*) Power spectra of the chip profiles clearly showing the periodic nature of the lower trace in (*a*) and the more aperiodic nature of the upper trace.

Brown, 1986) but not emphasized. Currently NIST researchers are testing the hypothesis that this transition in behavior is due to the nonlinear dynamics of the material deformations.

Dynamic models that might provide clues about the complexity of the observed behavior are discussed below.

Dynamic Models of Segmented Chip Formation

The primary mechanisms for dynamic interaction between catastrophic shear zones are elastoplastic loading and thermal diffusion. Figure 3.8 shows the formation of a segmented chip. Catastrophic shear was initiated along the line $B'D'$ (inclined at an angle θ relative to the workpiece surface) when the tool tip was at B'. As the tool tip moves from B' to A, the stress required to cause further deformation in shear zone BD decreases due to thermal softening, and the tool experiences rapid unloading along BE. Simultaneously the stresses

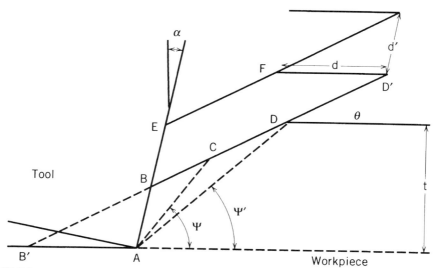

FIGURE 3.8. Schematic showing the formation of a chip segment during cutting with cutting speed V_c, uncut chip thickness t, rake angle α. The spacing between segments measured parallel to the tool rake face is denoted d, while the segment spacing measured along the original "free" surface of the workpiece is denoted d'.

increase along AB as the material in region $B'AB$ is locally indented and sheared. Heat generated during this indentation process and in the shear zone diffuses into the workpiece material ahead of the tool, lowering its yield strength. It is the combination of the thermal state of the workpiece ahead of the tool and the stresses applied along both AB and BD that determine when the next chip segment will form. This thermal and elastic interaction between the formation of one shear zone and the next provides the potential for dynamics in the system. These dynamics my be quite complicated due to strong nonlinearities in the constitutive behavior of the material. This complexity could lead to aperiodic variations in the spacing between shear zones similar to those observed experimentally. The dynamics of chip segmentation will have a direct effect on the cutting forces and will likely affect such practically important process parameters as rate of tool wear.

Although the above discussion about the formation of segmented chips is instructive, its complexity precludes analytical solution. For this reason most analytical treatments (Clifton et al., 1984; Bai, 1982; Burns, 1985; Wright and Walter, 1987) of shear band formation have focused on the simplified problem of one-dimensional shear in an elastoplastic solid. Figure 3.9a shows such a model in which a slab is assumed to undergo a simple one-dimensional shearing with thermal softening and heat conduction. The equations describing the deformation of this slab can be expressed as (Wright and Walter, 1987)

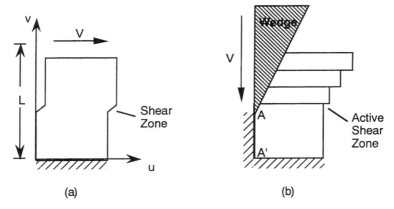

(a)   (b)

FIGURE 3.9. One-dimensional models of thermoplastic shear. (*a*) Deformation of a slab of material subjected to an average strain rate V/L. (*b*) Deformation of a slab of material subjected to stresses imposed by an impinging wedge moving downward with velocity V.

$$\rho \dot{u} = s_{,v} \tag{1a}$$

$$\dot{s} = \mu(u_{,v} - \dot{\gamma}_p), \tag{1b}$$

$$\rho c \dot{\theta} = (k\theta_{,v})_{,v} + s\dot{\gamma}_p, \tag{1c}$$

$$f(s, \theta, \dot{\gamma}_p) = \kappa, \tag{1d}$$

$$\dot{\kappa} = M\dot{\gamma}_p. \tag{1e}$$

Equation (1a) is a balance of linear momentum, where s is the shear stress, ρ is the material density, and $u(v)$ is the displacement of the material at the position v. The overdot represents the partial derivative with respect to time, and the term $(\ )_{,v}$ represents the partial derivative with respect to the position v. Equation (1b) states that the stress rate is proportional to the elastic strain rate, where $\dot{\gamma}_p$ is the plastic strain rate and μ is equal to the shear modulus of the material. Equation (1c) is a statement of thermal energy balance, where $\theta(v)$ is the temperature of the material as a function of v, and k and c are the conductivity and heat capacity of the material, respectively. Equation (1d) is the constitutive law relating the stress, strain rate, and temperature. Finally Equation (1e) prescribes the evolution of the yield strength κ as a function of plastic strain rate; in this case it is a simple linear strain hardening law. Equations of this form have been thoroughly analyzed using linear perturbation methods, asymptotic analysis and finite-element analysis (Clifton et al., 1984; Bai, 1982; Burns, 1985; Wright and Walter, 1987). These analyses show that the equations become unstable as the strain rate is increased, and this instability is manifested as a localization of strain. The exact behavior is found to be sensitive to the material constants and the form of the constitutive laws.

One of the major challenges in quantitative prediction of shear localization is the proper determination of these constants and the functional form of the constitutive laws.

Analysis of chip formation is more difficult than steady shear because the tool introduces a moving load. However, the essence of the problem is captured by considering the one-dimensional shear failure of a column of material under the action of an impinging wedge, as shown in Figure 3.9b. The material elements are assumed to be held fixed along line AA' until released by the wedge as it moves downward with velocity V. Although the equations describing the deformations in this system are Equations (1), the boundary conditions are more complicated and make the problem difficult to analyze. Davies et al. (1986) have done some preliminary work in which the partial differential equations for the system in Figure 3.9b have been discretized into a set of coupled nonlinear ode's, and their behavior was studied numerically. These simulations show some behavior that is qualitatively similar to that observed in the orthogonal cutting experiments described above. In addition Burns and Davies (1997) used a line of analysis similar to that used to explain oscillatory behavior in a low temperature torsion test. This approach has produced some analytical results for the stability of material flow in chip formation and the types of dynamics that might be expected in the flow.

3.2.3 Discussion

Hard-turning is a developing industrial process with numerous applications for manufacturing precision components that can resist wear (Soons and Yaniv, 1996). However, the hard-turning process is still limited by tool wear, residual stresses, and a lack of dynamic stability which limits attainable surface finish. There are two major technical areas where techniques of nonlinear dynamics might be applied to hard-turning. The first is to increase understanding of cutting-induced vibrations. This could help produce more effective algorithms for control and possibly provide techniques for recognizing tool wear. The second area is the understanding of material flow patterns which can become highly inhomogeneous as cutting speeds are increased. These flow patterns are governed by dynamic equations that are highly nonlinear due to the constituive behavior of the material. Understanding the dynamics of material flow during hard-turning could yield improvements in the prediction of cutting forces and possibly provide a link between materal flow and other important process variables such as the rate of tool wear.

3.3 HIGH-SPEED MILLING

The definition of high-speed milling is dependent on tool and workpiece materials. In general, "high speed" refers to cutting speeds that are significantly greater than those currently used in a particular process (King, 1985). The motivation to increase cutting speeds in machining is to realize higher material removal rates and reduce production time and production costs (Soons and Yaniv, 1996). Saloman was one of the first proponents of high-speed machining. From the period from 1924 to 1931, he collected experimental data on the variation of machining temperature with cutting speed. These data indicated that for a range of nonferrous materials, cutting temperature peaked and then declined as the cutting speed was increased. This suggested that more favorable cutting conditions could be attained by increasing cutting speeds well beyond those conventionally used at that time. Although researchers have not been able to reproduce Saloman's results, the suggestion that machinability can be improved or at least maintained while cutting speed is increased sparked significant interest in high-speed machining. Current research suggests that cutting temperatures do not decrease but instead plateau as cutting speed is increased (Gettelman, 1981) and that the primary factors that limit maximum practical cutting speeds are thermally induced tool wear and the availability and cost of high-speed machines. In practice, cutting speeds are chosen as a compromise between a number of competing physical and economic factors. An excellent history of the development of high-speed machining technology over the period from the early work of Saloman through the mid-1980s can be found in King (1985).

The physical causes of tool wear are numerous and include fracture, abrasion, adhesion, and chemical wear. These factors often compete and interact. At high cutting speeds, material temperatures increase and peak stresses decrease due to thermal softening of the material. Thermally activated wear processes are typically exponentially sensitive to temperatures at the tool-chip interface, roughly following an Arhenius-type activation law. As a result thermally activated wear mechanisms such as chemical reactions and diffusion tend to dominate over mechanical ones. Because these processes are also very sensitive to chemical composition, limiting speeds are a strong function of tool and workpiece materials. For example, research by McGee (Gettelman, 1981) suggests that as the speed is increased in machining aluminum, cutting temperatures plateau at the melting temperature (480°–600°C). At these temperatures reaction rates are not high enough to cause significant chemical wear in commercially available carbide, high-speed steel and ceramic tools. Thus the maximum speed for machining aluminum is limited by available machining technology. By contrast, titanium is chemically reactive at the temperatures generated in machining. As a result machining speeds for titanium alloys are limited by the availability of tool materials that resist chemical wear at elevated temperatures. Figure 3.10, after Shultz and Moriwaki (1992) and Soons and Yaniv (1996), further illustrates the variation in maximum attainable cutting speeds for various materials including aluminum and titanium.

Research suggesting that aluminum can be machined at very high speeds without significant increases in tool-wear rates has been a major driving factor behind recent industrial interest in high-speed milling. This interest has

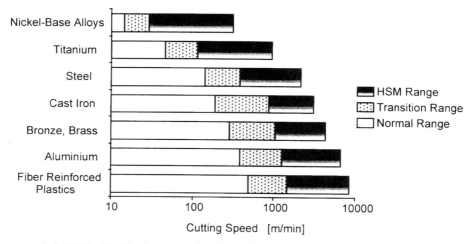

FIGURE 3.10. Maximum attainable cutting speeds for different materials.

prompted the development of reliable, computer controlled, high-speed spindle and slide technologies. Milling centers with maximum spindle speeds in excess of 40,000 rpm (666 Hz) and sustained power output of 40 kw are becoming commercially available. By using linear drive motors instead of lead screws, slides with maximum speeds exceeding 1 m/s and accelerations exceeding 1 g are feasible. With these capabilities, material removal rates in aluminum are high enough to justify machining parts that would previously have been forged or cast. In addition costs of assembly, fixture storage, and tracking can be substantially reduced by combining separate components into a single machined part. Important applications of high-speed milling include manufacture of thin-walled, ribbed, aluminum components for aircraft, steel tire molds, and automotive transmission cases.

Because of the high spindle speeds and high slide accelerations, maintaining accuracy in high-speed machining requires the development of more accurate thermal and dynamic models of the machine and the cutting process. One important area of research is the prediction and elimination of vibrations resulting from *dynamic tool-workpiece interactions*. In milling, these interactions are particularly difficult to predict because the intermittent cut introduces an *impact-type nonlinearity* (Holmes, 1982; Moon, 1992; Shaw and Shaw, 1989; Fang and Wickert, 1994). Impact-induced dynamics are of particular concern in high-speed milling where the geometry of machined parts often introduces significant flexibility in the tool and workpiece (Tlusty, 1993) and the cutter frequencies are high enough to interact with numerous vibration modes of the tool-workpiece system. A standard machinist's rule-of-thumb is that to maintain vibrational stability, the length-to-diameter ratio of a cutting tool should not exceed 5:1. This places significant limitations on the geometry of machined parts. If the vibrations induced by the intermittent cut were understood and controlled, the dynamic limitations of the process could be lessened, and manufactures would have greater freedom to design and machine more complex components. It is also worth noting that advances in technology are difficult to implement on the shop floor, as machinists are often reluctant to adopt new procedures (Soons and Yaniv, 1996).

Previous work on the prediction of vibrations in milling has focused on the phenomenon of self-excited oscillations that result from overcutting a previously cut surface (regenerative chatter). Hanna and Tobias (1974) use first-order harmonic balance to determine the conditions for steady state chatter for a single degree-of-freedom model of a milling operation with nonlinear stiffness and nonlinear time-delay terms. Tlusty and Ismail (1983) and Tlusty (1986) provide a more detailed analysis of milling stability that emphasizes the influence of "process damping" and the possibility of interactions between regenerative vibrations and vibrations induced by "the fundamental period of the process." Tlusty and Ismail (1983) acknowledged the importance of intermittency in milling dynamics. They developed a numerical simulation to examine these vibrations. However, the possibility that the intermittency itself could lead to vibrational instabilities was not discussed. Shridar et al. (1968a, 1968b) and Hohn et al. (1968) derived a model for the milling operation that includes time-dependent coefficients resulting from the complex geometry and intermittency inherent in the process. An analytical algorithm was developed and implemented on a computer, and the results of the stability calculation were found to agree well with experimental data. Minis and Yanushevsky (1993) derived a model similar to that of Shridar et al. (1968a). The stability of this model was then analyzed in the frequency domain. Although the intermittency of the milling operation is included in the models of Shridar et al. and Minis et al., the contact time between each tooth and the workpiece is assumed to be constant and known, leading to a periodic forcing term that is ignored in the stability calculations. A shortcoming of their approach is that the inherent impact nonlinearity, and the associated array of dynamic behavior and possible dynamic instabilities are ignored.

To make accurate predictions about process stability and the finish of machined surfaces in high-speed milling, the effect of impact nonlinearities must be explored. In the remainder of this section, experimental and numerical evidence supporting this assertion are presented.

3.3.1 Nonlinearities in Milling

In this section the derivation of a simplified model of the milling process, closely paralleling the derivations found in Shridar et al. (1968a) and Minis and Yanushevsky (1993), is presented. The different types of nonlinearities encountered during milling are discussed, with particular emphasis on the impact-type nonlinearity discussed above and its importance for the case of high-speed milling.

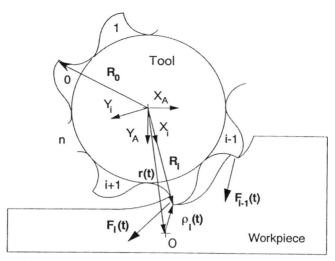

FIGURE 3.11. Schematic of a slab milling operation. A downmilling operation with an *n*-flute cutter is shown.

Figure 3.11 is a schematic of a slab milling operation where, for simplicity, only planar motions are considered. The cutter has n equally spaced teeth and rotates clockwise with an angular velocity of Ω cycles per second. The workpiece moves to the left with a constant feed rate f_0. The type of operation shown is referred to as a *down-* or *climb-milling* operation. Restricting the model to consider the machining of thin-walled components, the workpiece is assumed to be flexible and the cutting tool rigid, with its center remaining fixed in space. The intertial coordinate frame $\{A\}$ defined by axes (X_A, Y_A) is fixed at the center of rotation of the cutter. The vector $\mathbf{r}$ represents the position of a point, O, rigidly fixed to the workpiece, measured relative to the origin of coordinate frame $\{A\}$. The numbered coordinate frames $\{0\}, \{1\} \dots \{n\}$ defined by the vectors $(X_0, Y_0)(X_0, Y_0) \dots (X_n, Y_n)$ rotate with angular velocity Ω and remain fixed to the cutter with each X_i-axis lying along a line from the origin of $\{A\}$ to the tip of the ith cutter tooth. At time $t = 0$, coordinate frames $\{A\}$ and $\{0\}$ coincide so that the angle of the ith cutting tooth (measured positive clockwise) is given by $\theta_i = 2\pi(\Omega t + i/n)$. The force exerted by the ith tooth on the workpiece is denoted $\mathbf{F}_i$. The workpiece dynamics can be represented by differential equations in the following way:

$$^A\mathbf{r} = \mathbf{H}(D) \sum_{i=1}^{n} {}^A\mathbf{F}_i. \tag{2}$$

Here the notation $^A\mathbf{P}$ is used to denote the representation of the vector $\mathbf{P}$ with respect to the Cartesian coordinate frame $\{A\}$ and D is the differential operator

d/dt. If the workpiece dynamics are linear, $\mathbf{H}(D)$ is a two-by-two matrix

$$\mathbf{H}(D) = \begin{pmatrix} H_{xx}(D) & H_{xy}(D) \\ H_{yx}(D) & H_{yy}(D) \end{pmatrix} \tag{3}$$

that represents the dynamic response of the workpiece to forces applied at the point of contact with the cutting tool.[4]

There are numerous models for predicting the forces exerted by each cutting tooth on the workpiece. For example, Klein et al. (1982) develop a full three-dimensional model for predicting cutting forces from a multi-toothed cutter with a nonzero helix angle. In this case we will take a more simplified approach, assuming that each tooth performs an orthogonal cut, and that each tooth interacts only with the surface cut by the tooth immediately preceding it. Minis and Yanushevsky (1993) suggest a model of the form

$$^{i}\mathbf{F}_{i}(t) = h_{i}(\boldsymbol{\rho}_{i}(t), \boldsymbol{\rho}_{i+1}(t - \tau), t)b(C_{1}(D)^{i}\boldsymbol{\rho}_{i}(t) - C_{2}^{i}\boldsymbol{\rho}_{i+1}(t - \tau)), \tag{4}$$

where $\mathbf{F}_{i}$ is the cutting force exerted by the ith tooth written with respect to the rotating frame $\{i\}$, $\mathbf{R}_{i}$ is a vector representing the position of the cutting edge of tooth i, $\boldsymbol{\rho}_{i}(t) = \mathbf{r}(t) - \mathbf{R}_{i}(t)$ represent the relative distances between the workpiece center and the cutter teeth, b is the width of cut, $h_{i}(\boldsymbol{\rho}_{i}(t), \boldsymbol{\rho}_{i+1}(t - \tau), t)$ is a nonlinear function equal to 1 when the tooth i is in contact with the workpiece and equal to zero when it is not in contact, $\tau = 1/\Omega n$ is the time delay between the passage of consecutive cutter teeth, and $C_{1}(D)$ and C_{2} are two-by-two matrices containing the cutting force constants. Because of the assumption that each tooth interacts with the surface cut by the tooth preceding it, F_{i} depends only on the cutting history of tooth $i + 1$. In reality this may not be the case, and F_{i} may depend on the past history of other teeth as well.

One of the simplest models for cutting is to assume that the tangential component of the cutting force is linearly proportional to the depth of cut and that the radial component is proportional to the tangential component. In this case $C_{1}(D)$ and C_{2} are given by

$$C_{1} = C_{2} = \begin{pmatrix} k_{c} & 0 \\ \lambda k_{c} & 0 \end{pmatrix}, \tag{5}$$

where k_{c} is the specific cutting energy and λ is the proportionality constant for the radial force component. Linear damping can be added to the cutting model

---

[4]This representation is valid if all forces are applied at a single point on the workpiece. This is approximately true if the wavelengths of the important structural modes of the workpiece are much larger than the distance over which the cutter contacts the workpiece. Note that in reality $\mathbf{H}(D)$ changes slowly with time as the cutter moves across the workpiece. This effect is not considered here.

by adding a linear term in the operator D to the entries of the first column of C_1. For these special cases the cutting mechanics are assumed to be linear. This is not generally the case. For example, Klein et al. (1982) suggest that the specific cutting energy k_c is related to the chip thickness by a power law.

To complete the model, the representation of F_i must be transformed into the coordinate frame $\{A\}$ and substituted into Equation (2) as follows:

$$^A\mathbf{r} = \mathbf{H}(D) \sum_{i=1}^{n} {}_i^A Q^i \mathbf{F}_i, \tag{6}$$

where the transformation matrix ${}_i^A Q$ is defined by

$$^A_i Q = \begin{pmatrix} \cos \theta_i & -\sin \theta_i \\ \sin \theta_i & \cos \theta_i \end{pmatrix}. \tag{7}$$

Equations (4), (6), and (7) represent a nonlinear model of the milling operation containing time-dependent coefficients in the rotation matrix ${}_i^A Q$.

There are numerous possible sources of nonlinearity in milling. As pointed out above, Hanna and Tobias (1974) have examined the effect of nonlinearities in both the structural behavior of the machine and in the cutting mechanics. However, the effect of the impact nonlinearity inherent in the function $h_i(\boldsymbol{\rho}_i(t), \boldsymbol{\rho}_{i+1}(t-\tau), t)$ in Equation (4) has not been examined. This term is similar to nonlinearities known to dominate the dynamics of several simpler paradigm problems in nonlinear dynamics (Holmes, 1982; Shaw and Shaw, 1989; Moon, 1992) such as the bouncing of a ball on a sinusoidally vibrating table (Holmes, 1982). In studies of regenerative instabilities in milling, this term is assumed to be one for a constant and predetermined length of time. This assumption yields linear equations of motion. However, in general, the value of h_i depends on the position of the workpiece, the angle of the cutting tool, and the functional shape of the machined surface (past history of workpiece motions). The time of contact between each tooth and the workpiece may itself have a complex dynamic behavior, analogous to the *time-of-flight* defined by Holmes for the bouncing ball problem (Holmes, 1982). Impact nonlinearities can lead to numerous complex nonlinear phenomena including period-doubling bifurcations, chaotic dynamics, and other dynamic instabilities that could damage the workpiece, the tool or the machine. In high-speed milling operations, cutter rotation rates are high, and workpieces and tools are flexible. The number of flutes on the cutting tool typically does not exceed four, and impact dynamics can be expected to dominate vibrational behavior. In the next section the importance of impact nonlinearity in the milling of thin-walled parts is demonstrated.

3.3.2   Milling of Thin-Walled Parts — Experimental Results

A set of milling experiments were conducted on a conventional CNC machine located at NIST. The workpiece consisted of a 4.8-mm-thick, 50.8-mm-wide,

TABLE 3.2
Five Lowest Natural Frequencies of the Workpiece

Mode Number	Measured Modal Frequency (Hz)
1	416
2	1248
3	2112
4	2560
5	2944

95.3-mm-long aluminum beam. It was clamped in a vice at one end, and the free end was down-milled using a steel, two-flute, 12.7-mm-diameter cutter. Workpiece bending vibrations occurring during milling were measured using a pair of foil strain gages placed on opposite sides of the base of the beam and connected to a Wheatstone bridge, an analog filter, and a digital oscilloscope. The filtering was low pass with a cutoff at 5 kHz, and the data sample rate was 20 kHz. Data were then digitally filtered at 2 kHz before phase portraits were plotted. The first five natural frequencies of the beam were measured. The results are displayed in Table 3.2. The two lowest frequencies, 416 Hz and 1248 Hz, correspond to the first bending and torsional modes of the workpiece, respectively. The rotational speed of the cutter was varied from 500 rpm to 2000 rpm, and the radial depth of cut was varied from 0.13 mm to 0.76 mm. For all experiments the axial depth of cut was fixed at 4.8 mm, and the feed per tooth was held constant at 0.06 mm. Milling passes were made at the end of the workpiece in 25.4-mm-long passes centered on the workpiece width. Between cuts the workpiece was carefully clamped to eliminate vibrations, and the surface was milled flat. In this fashion the dynamics of each pass was independent of the surface produced by the previous cut. To further simplify the analysis, no coolant was used during cutting.

Figure 3.12 shows a voltage signal corresponding to the bending of the workpiece, taken when the cutter was near the center of its 25.4-mm-long cut.

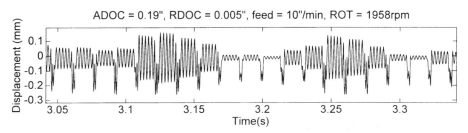

FIGURE 3.12. Time history of voltage produced by strain gage pair at the base of the workpiece.

For this example the radial depth of cut was 0.13 mm, and the cutter rotation rate Ω was approximately 32.6 revolutions per second (approximately 2000 rpm), corresponding to a cutting speed of approximately 1.3 ms^{-1}. There are several features of the time trace worth noting. First, the large negative excursions occurring at approximately 15 ms intervals are the engagement of the cutting teeth with the workpiece. These are separated by intervals of lightly damped, free vibrations during periods when the workpiece is not engaged with the cutting tool. These free vibrations have a frequency that corresponds to the first mode of the workpiece. The amplitude of these vibrations appears to vary nonperiodically between contacts with the cutter.

The power spectrum of the motions is shown in Figure 3.13. This spectrum has a structure typical of all milling vibrations recorded in the series of experiments conducted at NIST. It has a set of harmonic peaks at multiples of the cutting tooth engagement frequency 66.2 Hz, a set of small subharmonic peaks at 32.6 Hz multiples, and a broad peak (or mound) between 350 and 450 Hz, which is assumed to correspond to the first-mode natural frequency of the workpiece.

The nonperiodic character of the vibrations is manifested in phase portraits of the milling dynamics. A set of phase portraits illustrating the change in character of the motions as the radial depth of cut was varied from 0.13 to 0.78 mm is shown in Figure 3.14. The phase portraits were created by numerically differentiating the time traces after the collection of the data. Each of the portraits has a space-filling character, indicating that the motions are not periodic for any of the radial depths of cut chosen. The set of concentric ellipses on the right of each portrait represents the free lightly damped motions of the workpiece between cutter engagements, and the excursions to the left represent the cutter engagements with the workpiece. Figure 3.15 shows two more phase portraits taken with radial depth of cut of 0.13 mm at 500 and 1000 rpm, respectively. The portrait from the 1000-rpm cut has a space-filling nonperiodic character, like the phase portraits discussed above. However, the phase portrait from the 500-rpm cut has a much more periodic character. The surface finish of the workpiece after the 500-rpm cut was found to be significantly better than that obtained at either 1000 or 2000 rpm. These data suggest that even for this very flexible workpiece, parameter values can be optimized to produce improved surface finish.

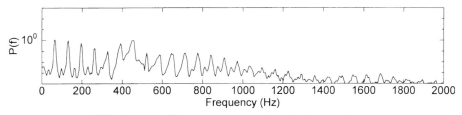

FIGURE 3.13. Power spectrum of workpiece motions.

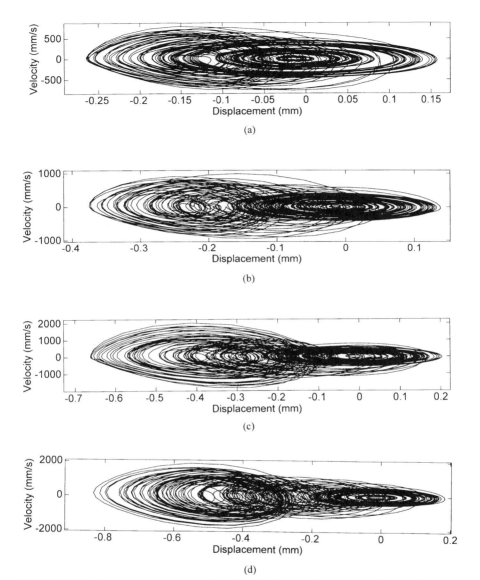

FIGURE 3.14. Example phase portraits of workpiece motions for radial depths of cut (a) 0.13 mm, (b) 0.26 mm, (c) 0.52 mm, and (d) 0.78 mm.

The method of delays (Nayfeh and Balachandran, 1995) was used to recreate the phase space of each data set with the delay chosen based on the characteristics of the autocorrelation function. From the delay space reconstruction the pointwise dimension of the attractors was calculated and found to be fractal. The results also indicate the presence of two scaling regions in

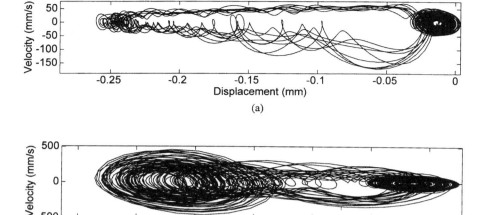

FIGURE 3.15. Phase portraits of workpiece motions for radial depths of cut of 0.13 mm and cutter rotations rates of (*a*) 500 rpm and (*b*) 1000 rpm.

several of the data sets. One region has a dimension consistently between 1 and 2, and the other between 4.2 and 1.7, depending on the cutting conditions. The lower-dimensional region probably corresponds to the free vibrations of the workpiece where two state variables are sufficient to describe the motions. This region appears as the set of concentric (slowly decaying) elliptical curves in the phase plane diagrams shown in Figures 3.14 and 3.15. The higher-dimensional region corresponds to the cutting interactions. During these interactions, the motions depend on the state of the workpiece, the position of the cutter (time), and the shape of the previously machined surface. The space-filling nature of the phase portraits, the decay of the autocorrelation of the data, and the fractal dimension of the attractors provide strong evidence that the milling dynamics are chaotic. The details of the data analysis summarized here can be found in Davies and Balachandran (1996b).

3.3.3   Impact Models

Modeling the dynamics of milling operations is extremely difficult because of the presence of impact nonlinearity, delayed forcing terms, and time-dependent coefficients. Nevertheless, a simplified impact model with properly chosen parameters can yield predictions that are consistent with experimental data. Figure 3.16 shows a model developed at NIST and its relation to a schematic of the milling process. The structural dynamics of the workpiece are represented by a single-mode, lumped parameter model with mass m, stiffness k_1, and damping c_1. The cutter motion is represented by a rigid impactor that moves

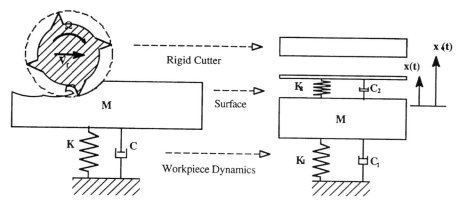

FIGURE 3.16. Impact model of milling process where the workpiece motions are represented with a single mode, the cutter is a rigid impactor, and the cutting interaction is represented with a linear stiffness and linear damping.

periodically to engage and disengage the workpiece. It is assumed that the interactions between the cutter and the workpiece can be approximated by a zone around the workpiece with stiffness k_2 and damping c_2. When the cutter impinges on this zone, the workpiece experiences additional elastic and damping forces proportional to its position and velocity relative to the cutting tool. For small depths of cut, k_2 is approximately the radial cutting stiffness. If we assume that the tangential cutting stiffness is $k_c b$, where k_c is the specific cutting energy and b is the width of cut, and that the radial cutting force is proportional to the tangential cutting force, then k_2 can be approximated by $\lambda k_c b$, where λ is the constant of proportionality.

The equations of motion for this model are

$$\dot{x} = v,$$

$$\dot{v} = \begin{cases} -\dfrac{1}{m}\left(k_1 x + k_2(x - x_0) + c_1 v + c_2\left(v - \dfrac{dx_0}{dt}\right)\right), & x > x_0, \\[4mm] -\dfrac{1}{m}(k_1 x + c_1 v) & x \geqslant x_0, \end{cases} \tag{8}$$

where the overdot represents differentiation with respect to time t, x is the position of the oscillator, v is the velocity of the oscillator, and x_0 is the position of the rigid impactor (cutting tool).

For a two-flute cutter the input excitation can be represented by a rectified periodic signal as

$$x_0 = A - B\cos(2\pi\Omega t)|, \tag{9}$$

where Ω is the rotation rate of the cutter in hertz, and $B - A$ is the penetration of the rigid impactor below the equilibrium position of the oscillator. In the milling experiments this parameter corresponds to the radial depth of cut. Equations (8) and (9) describe a bilinear oscillator similar to that first studied by Holmes and Shaw (Holmes, 1982). The equations are solvable in a piecewise manner as the impactor contacts and releases the oscillator. These solutions can be used to analytically determine the stability of various periodic motions of the system.

In simulations of the model described above, the parameters were given the following values: $m = 0.015\,kg$, $k_1 = 111\,kN/m$, $c_1 = 0.08\,N{\cdot}s/m$, $k_2 = 3.8\,MN/m$, and $c_2 = 5.0\,N{\cdot}s/m$. The values for m, k_1, and c_1 were chosen to represent the experimentally measured first-mode behavior of the workpiece. The cutting stiffness k_2 was calculated by taking $\lambda = 0.8$, $b = 4.8$ mm, the axial depth of cut, and $k_c = 1\,kN/mm^2$ (Kalpakjian, 1991), the approximate specific cutting energy for aluminum. Little experimental data are available on the value of the cutting damping c_2. The value given above is simply a "guess" that must be refined as more experimental data are collected. However, the value chosen suffices to show the relationship between the predicted and measured dynamics.

Figure 3.17 is a bifurcation diagram of the simulated dynamics of the model where the bifurcation parameter is the "depth of cut" $B - A$. This parameter is varied from 0.13 to 0.9 mm. The diagram was created by sampling the displacement of the oscillator at the points of maximum penetration of the rigid impactor; corresponding to sampling the position of the workpiece at the passage of each cutter tooth. Figure 3.17 shows a wide range of periodic and nonperiodic behavior. At 0.13-mm depth of cut, the motion has a period equal to that of the rigid impactor. A phase plane of this motion is shown in Figure 3.18. Clearly this motion does not correspond to the less regular motion of the experiment depicted in Figure 3.14a. However, at a depth of cut of 0.52 mm,

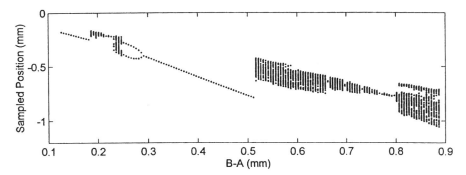

FIGURE 3.17. Bifurcation diagram of impact model of milling processes. The parameter $B - A$ is varied from 0.13 mm to 0.9 mm, while the excitation frequency is held constant at 32.6 Hz.

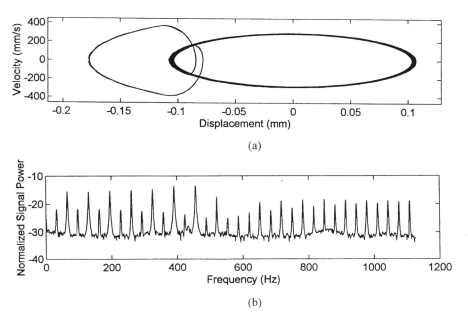

FIGURE 3.18. (*a*) Phase portrait and (*b*) power spectrum of model dynamics for $B - A = 0.13$ mm.

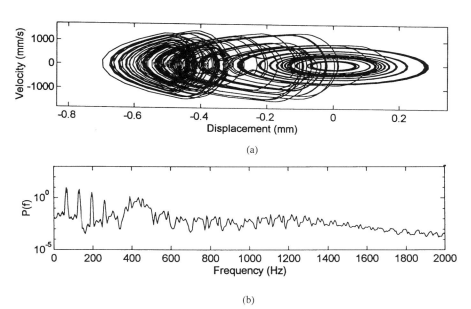

FIGURE 3.19. (*a*) Phase portrait and (*b*) power spectrum of model dynamics for $B - A = 0.52$ mm.

the model predicts a chaotic motion of the workpiece. A comparison of Figure 3.19 and Figure 3.14*b* shows a notable qualitative similarity between the measured and predicted phase portraits.

Throughout a series of milling experiments with a variety of different workpieces, few periodic motions have been observed. One notable exception is the motion depicted in Figure 3.15*a*, which appears to be nearly a closed curve in the phase plane. The impact model predicts a rich array of both periodic and nonperiodic motions. It is hypothesized that the relative abundance of nonperiodic motions observed in the experiments described above is the result of the changing machined surface, which adds degrees of freedom to the dynamics of the system. Work is currently underway at NIST to understand this effect.

3.3.4 Discussion

Numerous types of nonlinearities can effect the dynamic behavior of milling processes. These include nonlinearities in the cutting mechanics, nonlinearities in the machine behavior, and impact nonlinearities arising from the intermittent nature of the cut. Although the effects of the cutting and machine nonlinearities have received some attention, the effect of impact nonlinearities have been virtually ignored in the literature. Impact nonlinearities are likely to effect the stability of many milling operations with inherent workpiece and tool flexibility, high cutter rotation rates, and cutting tools with few flutes. Because of the strongly nonlinear nature of these motions, and evidence that many milling vibrations may actually be chaotic, new schemes controlling chaotic vibrations (Nayfeh and Balachandran, 1995) might be applicable. To implement such control schemes, new actuators capable of providing high-bandwidth variations in cutter position and/or feed rate would have to be developed. Successful implementation of vibration control would add an increased degree of flexibility for manufactures and designers utilizing high-speed milling processes.

3.4 HIGH-SPEED GRINDING

Grinding and other abrasive processes are among the oldest forms of machining, with their origins dating back to prehistoric times. Today these processes account for approximately 20–25% of total global expenditures on machining (Malkin, 1989). This relatively wide usage stems from both their unrivaled precision and their ability to machine hard and brittle materials such as hardened steels, glass, and ceramics. However, because material removal rates are typically lower than in milling and turning operations, typically grinding is only used in finishing operations. In an effort to increase material removal rates in grinding, high-speed grinding machines have been developed. These machines are capable of taking large depths of cut with low to moderate feed rates and very high cutting speeds sometimes in excess of 200 m/s. Wheels for these machines consist of so-called superabrasive materials such as CBN and diamond.

Machines for high-speed grinding and high-speed milling have been developing in parallel. Similar problems in the control of high-speed spindle and slide motions occur. However, in grinding static and dynamic cutting loads can far exceed those encountered in milling. Moreover the cutting wheel generally has a much larger radius and mass than a milling tool. To control static and dynamic deflections, grinding machines must be made much stiffer than their milling counterparts, thus increasing their cost. Malkin (1989) provides an excellent scientific and technical introduction to grinding processes as well as an extensive list of references on work done to date.

Modeling the dynamics of grinding operations is particularly intriguing because, unlike other machining processes, it involves significant dynamic variations in the shape of both the workpiece and the grinding wheel. Wear of the wheel is necessary to expose new abrasive grits; however, it may also be responsible for regenerative instabilities. To further complicate the problem, many grinding operations involve a driven motion of both the tool and the workpiece. This introduces a second delayed forcing term into the equations of motion of the system, rendering even the linear stability of the system very difficult to calculate.

Hahn (1954) was the first to recognize the importance of regenerative effects in grinding. He studied the regenerative stability of a linear model that included only workpiece regenerative effects. Following Hahn's work, several other researchers have studied regenerative chatter in grinding, focusing on single regenerative effects from either the wheel or the workpiece (Bartulucci and Lisini, 1969; Snoeys and Brown, 1969; Srinivasin, 1982). Snoeys and Brown developed a set of linear equations encompassing both wheel and workpiece regenerative effects. A similar model containing a single dynamic degree-of-freedom and two delayed forcing terms has been analyzed extensively by Hahn and Thompson (1977) and Thompson (1974, 1986a, 1986b, 1992). Most of these studies focus on predicting stability boundaries for grinding operations. However, as suggested by several researchers (see chapter 18 in King, 1985; Thompson, 1986a, Strinivasan, 1982; Snoeys and Brown, 1969), *most practical grinding processes are unstable*; thus it is important to understand their behavior both before and after the onset of instability. With this aim a few linear models for predicting the exponential growth rate of chatter in grinding have been developed (Thompson, 1986a, 1986b). Predicting this growth rate is important because it determines how frequently the grinding wheel must be dressed during machining. None of the models to date have included process nonlinearities in the calculation of chatter growth rates or their possible effect on limiting chatter amplitudes.

The dynamical issues associated with grinding are illustrated in Figure 3.20. This figure provides a schematic of a cylindrical plunge grinding operation. In this type of operation, the grinding wheel rotates in a clockwise direction with

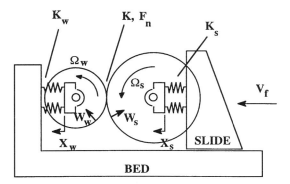

FIGURE 3.20. Cylindrical grinding operation.

angular speed Ω_s, while the workpiece rotates in a counterclockwise direction with angular speed Ω_w. The grinding wheel is brought into contact with the workpiece, and material is removed by the abrasive interaction between them. In a stable, steady-state operation, the wheel is fed toward the workpiece with a speed V_f. After the wheel and the workpiece make contact, a force F_n develops between them. This force is opposed by elastic forces in the machine structure as it deforms. Constants K_s and K_w represent the stiffnesses of the machine-tool structure, and X_w and X_s represent deflections. The rates at which the material is removed from the wheel and workpiece is a function of the normal force acting between them. Thompson (1974) developed a model based on this cylindrical grinding geometry. The key assumptions of Thompson's models are as follows:

1. Workpiece stiffness K_w and contact stiffness K are infinite.
2. Wheel and workpiece wear rates are linearly proportional to the normal force F_n.
3. Workpiece dynamics are linear.
4. The contact area between the workpiece and wheel is small.

These assumptions yield to a single degree-of-freedom linear model with two delayed forcing terms, one with delay $1/\Omega_w$ (workpiece regeneration) and the other with delay $1/\Omega_s$ (wheel regeneration). By assuming a periodic solution to the equations, Thompson calculated the stability boundaries (Thompson, 1974) for the process as a function of wheel and workpiece rotation rates. In a later set of papers (Thompson, 1986a, 1986b), Thompson extended this theory to study the dynamics of grinding after the onset of instability.

There are numerous nonlinearities inherent in the grinding process that have not been accounted for in past research. Because these nonlinearities may have a significant effect on the growth rate of chatter in grinding, understanding their sources and effects is of practical importance. Some nonlinearities encountered in grinding include:

- Hertzian contact between the wheel and the workpiece.
- Relationship between wheel and workpiece wear as a function of normal force (King, 1985).
- Large machine deflections induced by relatively high contact and cutting forces.
- Hydrodynamic effects of coolant flowing between wheel and workpiece at high speeds.
- Transitions in material removal mechanisms (for example from ductile behavior to brittle fracture).

Although these nonlinearities may have a substantial effect on the growth of instabilities in grinding, they are virtually ignored in the literature.

3.5 CONCLUSIONS

From the specific examples discussed in this chapter, several more general areas where the tools of modern nonlinear dynamics can be applied to improve our understanding of manufacturing processes can be identified. These include:

- The effect of nonlinearities on the dynamics of systems with delayed forcing.
- The dynamics of multiple-mode systems with delayed forcing.
- The growth of instabilities in nonlinear systems with two-delayed forcing terms.
- The control of spindle dynamics at high rotation rates (exceeding 30,000 rpm).
- The control of slide motions during high-speed (exceeding 1 m/s), high-acceleration (1 g) contouring motions with micrometer-level positioning accuracy, in the presence of nonlinearities such as friction and backlash.
- The development of simple nonlinear models to describe material flow instabilities.

Except as noted above, by and large dynamical issues, especially the effect of nonlinearities, have been ignored in manufacturing research. Several reasons might be offered to explain this:

- Manufacturing problems are complex. They require cooperation among researchers from a number of different disciplines, and models are often not amenable to conventional analytical techniques.

- Until the recent trend toward high-speed machining, most practical manufacturing operations were carried out in parameter regimes that were reasonably approximated as quasi-static. Manufacturing was treated as an art form. Any dynamical problems that did arise were solved on the shop floor by experienced machinists, through trial and error methods.
- Manufacturing problems are highly nonlinear; until the last few decades, techniques for analyzing their behavior did not exist.

The development of computer controllers and higher-speed machines, and the need for rapid, flawless production of complex components, have forced researchers to begin transforming manufacturing from an art form into a science. Recently developed analytical, numerical, and experimental techniques for understanding nonlinear dynamic systems promise to play a significant role in this transformation.

REFERENCES

Abler, J., Packman, A. B., and Akerley, S. W. 1995. Control of chatter in a vertical lathe. Presentation at Workshop on Nonlinear Dynamics in Material Processing and Manufacturing, San Diego, CA.

Anand, L. 1984. Some experimental evidence on localized shear bands in plane-strain. *Scripta Metall.* **18**:423–427.

Asaro, R. J., and Needleman, A. 1984. Flow localization in strain hardening crystalline solids. *Scripta Metall.* **18**:429–435.

Bai, Y. L. 1982. Thermo-plastic instability in simple shear. *J. Mech. Phys. Sol.* **30**:195.

Bartalucci, B., and Lisini, G. G. 1969. Grinding process instability. *J. Eng. Ind.* **91**:597–606.

Burns, T. J. 1985. Approximate linear stability analysis of a model of adiabatic shear band formation. *Quart. Appl. Math.* **43**:65–84.

Burns, T. J., and Davies, M. A. 1997. A nonlinear dynamics model for chip segmentation in machining. *Phys. Rev. Lett.*, forthcoming.

Burns, T. J. 1994. A simple criterion for the onset of discontinuous plastic deformation in metals at very low temperatures. *J. Mech. Phys. Sol.* **42**:797–811.

Chou, Y., and Evans, C. J. 1996. Microstructural effects in precision hard turning. In *1996 Int. Mech. Eng. Congr. Expo.*, November.

Clifton, R. J., Duffy, J., and Hartley, K. A. 1984. On critical conditions for shear band formation at high strain rates. *Scripta Metal.* **18**:443–448.

Costin, L. S., Crisman, E. E., Hawley, R. H., and Duffy, J. 1979. On localisation of plastic flow in mild steel tubes under dynamic torsional loading. *Institute of Physics Conf.*, no. 47, Oxford, pp. 90–100.

Davies, M. A., and Balachandran, B. 1996b. Nonlinear dynamics in the milling of thin-walled structures. *J. Sound Vib.*, forthcoming.

Davies, M. A., Evans, C. J., and Harper, K. K. 1995. Chip segmentation in machining AISI 52100 steel. In *ASPE 11th Ann. Meet*, p. 235–238.

Davies, M. A., Chou, Y., and Evans, C. J. 1996a. On chip morphology, tool wear and cutting mechanics in finish hard turning. *Ann. CIRP* **45**:77–82.

Davies, M. A., Burns, T. J., and Evans, C. J. 1997. On the dynamics of chip formation in hardened materials. *Ann. CIRP* **46**, forthcoming.

Fang, F., and Wickert, J. A. 1994. Response of a periodically driven impact oscillator. *J. Sound Vib.* **170**:397–409.

Feeny, B. F. 1989. Autocorrelation and symbol dynamics for a dry friction oscillator. *Phys. Lett. A*, **141**:397–400.

Gettelman, K. 1981. High-speed milling: Where do we stand? *Mod. Mach. Shop*, February 1981.

Glimm, J., Plohr, B. J., and Sharp, D. H. 1993. A conservative formulation for large deformation plasticity. *Appl. Mech. Rev.* **46**:519–526.

Hahn, R. S. 1954. On the theory of regenerative chatter in precision grinding operations. *Trans. ASME* **76**:593–597.

Hahn, R. S., and Thompson, R. A. 1977. On the doubly regenerative stability of a grinder: The effect of wheel and workpiece speed. *J. Eng. Ind.* **99**:921–923.

Hanna, N. H., and Tobias, S. A. 1974. A theory of nonlinear regenerative chatter. *J. Eng. Ind.* **96**:247–255.

Hartley, K. A., Duffy, J., and Hawley, R. H. 1987. Measurement of the temperature profile during shear band formation in steels deforming at high strain rates. *J. Mech. Phys. Sol.* **35**:283–301.

Hohn, R. E., Shridar, R., and Long, G. W. 1968. A stability algorithm for a special case of the milling process. *J. Eng. Ind.* **90**:326–329.

Holmes, P. J. 1982. The dynamics of repeated impacts with a sinusoidally vibrating table. *J. Sound Vib.* **84**:173–189.

Kalpakjian, S. 1991. *Manufacturing Processes for Engineering Materials*. Addison-Wesley, Reading, MA.

King, R. I. 1985. *Handbook of High-Speed Machining*. Chapman and Hall, London.

Klein, R. G., and Nachtigal, C. L. 1975a. A theoretical basis for active control to improve boring bar operation. *J. Dynamic Systs. Meas. Control* **97**:172–178.

Klein, R. G., and Nachtigal, C. L. 1975b. The application of active control to improve boring bar performance. *J. Dynamic Systs. Meas. Control* **97**:179–183.

Klein, W. A., Devor, R. E., and Lindberg, J. R. 1982. The prediction of cutting forces in milling with applications to cornering cuts. *Int. J. Mach. Tool Des. Res.* **22**:7–22.

Komanduri, R., and Brown, R. H. 1981. On the mechanics of chip segmentation in machining. *J. Eng. Ind.* **103**:33–50.

Komanduri, R., and Brown, R. H. 1986. On shear instability in machining a nickel iron alloy. *J. Eng. Ind.*, **108**:93–100.

Komanduri, R., Schroeder, T., Hazra, J., and von Turkovich, B. F. 1982. On the catastrophic shear instability in high-speed machining of an AISI 4340 Steel. *J. Eng. Ind.* **104**:121–131.

Konig, W., Berktold, A., and Koch, K.-F. 1993. Turning versus grinding—A comparison of surface integrity aspects and attainable accuracies. *Ann. CIRP*. **42**:39–43.

Konig, W., Klinger, M., and Link, R. 1980. Machining of hard materials with geometrically defined cutting edges. *Ann. CIRP*: 61–64.

Litonski, J. 1977. Plastic flow of a tube under adiabatic torsion. *Bull. Acad. Pol. Sci., Serie Sci. Tech.* **25**:1–8.

Malkin, S. 1989. *Grinding Technology.* Ellis Horwood, Chichester.

Marchand, A., and Duffy, J. 1988. An experimental study of the formation of adiabatic shear bands in a structural steel. *J. Mech. Phys. Sol.* **36**:251–283.

Marusich, T. D. and Ortiz, M. 1995a. A parametric finite-element study of orthogonal high-speed machining. Submitted to *J. Eng. Ind.*, forthcoming.

Marusich, T. D., and Ortiz, M. 1995b. Modeling and simulation of high-speed machining. *Int. J. Num. Methods Eng.* **38**:3675–3694.

Mason, F. 1992. Hard turning is not a black art. *Amer. Mach.* (March): 41–43.

Minis, I., and Yanushevsky, R. 1993. A new theoretical approach for the prediction of chatter in milling. *J. Eng. Ind.* **115**:1–8.

Molinari, A., and Clifton, R. J. 1987. Analytical characterization of shear localization in thermoviscoplastic materials. *J. Appl. Mech.* **54**:806–812.

Moon, F. C. 1992. *Chaotic and Fractal Dynamics.* Wiley, New York.

Nachtigal, C. L., and Cook, N. H. 1970. Active control of machine tool chatter. *J. Basic Eng.* **92**:238–244.

Nayfeh, A. H., and Balachandran, B. 1995. *Applied Nonlinear Dynamics.* Wiley, New York.

Noaker, P. M. 1992. Hard facts on hard turning. *Manuf. Eng.* **108**: 43–46.

Olmstead, W. S., Nemat-Nasser, S., and Ni, L. 1994. Shear bands as surfaces of discontinuity. *J. Mech. Phys. Sol.* **42**:697–709.

Ramalingam, S., and Black, J. T. 1973. An electron microscopy study of chip formation. *Metall. Trans.* **4**:1103–1112.

Recht, R. F. 1964. Catastrophic thermoplastic shear. *J. Appl. Mech.* **31**:189–193.

Semiatin, L., and Rao, S. B. 1983. Shear localisation during metal cutting. *Mat. Sci. Eng.* **61**:192.

Shaw, J., and Shaw, S. W. 1989. The onset of chaos in a two-degree-of-freedom impact oscillator. *J. Appl. Mech.* **56**:168–174.

Shaw, M. C. 1993. Chip formation in the machining of hardened steel. *Ann. CIRP* **42**:29–31.

Shiriashi, M., Yamanaka, K., and Fujita, H. 1991. Optimal control of chatter in turning. *Int. J. Mach. Tools Manuf.* **31**:31–43.

Shridar, R., Hohn, R. E., and Long, G. W. 1968a. A general formulation of the milling process equation. *J. Eng. Ind.* **90**:317–324.

Shridar, R., Hohn, R. E., and Long, G. W. 1968b. A stability algorithm for the general milling process. *J. Eng. Ind.* **90**:330–334.

Shultz, H., and Moriwaki, T. 1992. High-speed machining. *Ann. CIRP* **42**:637–643.

Soons, H. A., and Yaniv, S. 1996. Precision in machining: Research challenges. National Institute of Standards and Technology, Internal Report (NISTIR 5628), Gaithersburg, MD.

Snoeys, R. and Brown, D. 1969. Dominating parameters in grinding wheel and workpiece stability. In *10th Int. MTDR Conf.*, pp. 325–348.

Srinivasan, K. Application of the regeneration spectrum method to wheel regenerative chatter in grinding. *J. Eng. Ind.* **104**:46–53.

Tewani, S. G., Rouch, K. E., and Walcott, B. L. 1995. A study of cutting process stability of a boring bar with active dynamic absorber. *Int. J. Mach. Tools Manuf.* **35**:91–108.

Thompson, R. A. 1974. On the doubly regenerative instability of a grinder. *J. Eng. Ind.* 96:275–280

Thompson, R. A. 1986a. On the doubly regenerative stability of a grinder: The theory of chatter growth. *J. Eng. Ind.* **108**:75–82.

Thompson, R. A. 1986b. On the doubly regenerative stability of a grinder: The mathematical analysis of chatter growth. *J. Eng. Ind.* **108**:83–91.

Thompson, R. A. 1992. On the doubly regenerative stability of a grinder. The effect of contact stiffness and wave filtering. *Trans. ASME* **114**:53–60.

Tlusty, J. 1986. Dynamics of high-speed milling. *J. Eng. Ind.* **108**:59–67.

Tlusty, J. 1993. High-speed machining. *Ann. CIRP* **42**:733–738.

Tlusty, J., and Ismail, F. 1983. Special aspect of chatter in milling. *ASME J. Vib., Stress, Reliab. Des.* **105**:24–32.

von Turkovich, B. F. 1971. ASTME Technical Paper No. MR-71-903.

Wright, T. W., and Walter, J. W. 1987. On stress collapse in adiabatic shear bands. *J. Mech. Phys. Sol.* **35**:701–720.

Zener, C., and Hollomon, J. H. 1944. Effect of strain rate upon plastic flow in steel. *J. Appl. Phys.* **15**:22–32.

Zhen-Bin, H., and Komanduri, R. 1995. On a thermomechanical model of shear instability in machining. *Ann. CIRP* **44**:69–73.

4

 (chapter rule with arrows)

CHATTER DYNAMICS IN SHEET-ROLLING PROCESSES

R. E. JOHNSON and H. P. CHERUKURI

4.1 INTRODUCTION

A commonly encountered problem in the cold-rolling of thin sheets in tandem rolling mills is the mechanical vibrations of the rolls. These vibrations often lead to unacceptable gage variations (often as much as 50%), striations on the rolled-strip, undesirable operating conditions and damage to the mill components. Depending on the frequency range in which they occur, three types of vibrations, namely, torsional oscillations, third-octave and fifth-octave chatter modes have been observed. The torsional mode falls in the frequency range of 5–15 Hz, the third-octave chatter lies in the range 125–240 Hz and the fifth-octave chatter, in the range 550–650 Hz. In the present chapter, we focus on the third-ocatave and fifth-octave chatter modes that cause vertical vibrations of the backup and work-rolls. Various factors such as high interstand tensions, thin gage strips, degrading lubricants, and high roll-speeds can cause these vertical vibrations, and this chapter aims to understand the role played by some of these factors.

Several theoretical models have been proposed to study the causes of chatter, and almost all of these models are based on the theory of multiple degree-of-freedom vibrating systems, that is, that the elastic deformation between the back-up rolls and the work-rolls provides a springlike restoring force. This restoring force associated with the contact deformation at the interfaces between the rolls is inherently nonlinear. The plastic deformation of

Dynamics and Chaos in Manufacturing Processes, Edited by Francis C. Moon.
ISBN 0-471-15293-5 © 1998 John Wiley & Sons, Inc.

the roll-bite between the work-rolls provides an additional force influencing the chatter motion of the rolls, and a nonlinear damping force can be supplied by various factors such as variable interstand tension and the lubricant state at the roll-sheet interface. When the damping force is positive, energy is dissipated out of the system and the mill-roll vibrations die out eventually. However, when appropriate conditions involving a combination of factors such as roll-speed, friction coefficient, strip tension, strip width, and reduction are met, the effective damping and the spring coefficients of the vibrating system can become negative. Consequently a self-sustained vibratory motion leading to a buildup of the vibration amplitude begins to take place until the effective spring and damping coefficients become positive again.

Different authors have considered these nonlinearities to varying degrees of approximation in studying the chatter motion of the mill-rolls. For example, the nonlinear force-displacement relation for the work–backup roll interaction is often linearized to make the analysis feasible. The plastic deformation of the sheet in the roll-bite between the rolls is often considered in the spirit of the classical slab-analysis approach due to Orowan. Also the chatter vibrations are often studied by considering vibration systems of various complexities. For example, if it is assumed that only the work-rolls are free to vibrate vertically and that the backup rolls are fixed, a 2-DOF (degree-of-freedom) vibrating system is obtained. On the other hand, when one considers both the backup and work-rolls to be vibrating, a 4-DOF vibrating system is obtained.

This chapter focuses mainly on the work of Johnson (1994) and Johnson and Qi (1994). However, we first present a brief review of existing studies pertaining to chatter in rolling-mills. This is followed by a detailed study of the role played by the inelastic deformation of the sheet in the chatter of rolling-mills. The friction condition between the work-rolls and the sheet is assumed to be proportional to the relative velocity between the work-rolls and the sheet. Further the sheet material is rate dependent and assumed to obey a power-law-type rigid-viscoplastic flow rule. The vertical motion of the rolls is decomposed into steady and dynamic parts. The steady part is associated with the nominal displacement of the work-rolls due to the large pressure in the roll-bite during steady state rolling of the sheet, whereas the dynamic part is associated with the chatter motion. The chatter motion causes gap variations between the work-rolls and hence additional plastic deformation in the sheet. This additional plastic deformation of the sheet, in turn, leads to the contribution of a dynamic force component to the total force on the rolls. The vertical motion of the rolls due to this force is approximated by simple multiple-DOF vibrating systems. Both 2-DOF and 4-DOF systems are considered in detail.

The roll-forces are obtained by means of regular perturbations where the ratio of the incoming sheet gauge or thickness to the roll-bite length is assumed small. The solution to the differential equations governing the chatter (roll-vibrations) is then obtained through the means of both linearization and asymptotic methods. The nonlinear force-displacement relation for the rolls is linearized, and the resulting dynamics is studied in detail. Analytic expressions for the natural frequencies are obtained, and the possibility of beat phenomena is also explored within the realms of this linear theory. Further the effect of the nonlinearity in the force-displacement relation for the 2-DOF system is studied through the Poincare-Lindstedt method.

It is also worth pointing out here that interstand tension has been proposed by many researchers as a major factor in causing chatter in multistand roll-mill systems. The mathematics of analyzing the interstand tension role in chatter can be tedious, and therefore this issue is not addressed here. Instead, we focus primarily on the role played by the plastic deformation of the sheet in the roll-bite in causing chatter. It has also been proposed that the chatter marks are primarily due to the bending mode vibrations of the rolls. This work approaches the vibrations of the rolls from the point of view of two-dimensional rigid-body vibrations and therefore precludes the consideration of bending modes.

4.2   LITERATURE REVIEW

Identification of chatter sources and prevention has led to the development of mathematical models aimed at predicting chatter sources and describing chatter phenomenon. In the following we review some of the mathematical models proposed to analyze chatter in rolling-mills. For a comprehensive survey on this subject, the reader is referred to Roberts (1988).

Roberts (1978) recognized that the elastic contact between the work and backup rolls provides a springlike restoring force and that the vibrations of the work-rolls when the backup rolls are relatively stable can be modeled by a mass-spring system with 1-DOF and the work-roll constituting the mass. Since the displacement-force relation between the work and backup rolls is nonlinear, the spring is nonlinear. However, guided by the observation that the work-roll displacements are extremely small, Roberts linearized this relation and provided an analytic expression for the fifth-octave mode clatter. Tamiya et al. (1977) have considerd a 4-DOF system and concluded that chatter is a self-excited vibration originating from the phase difference between the vertical vibration of the roll and the fluctuation of the entry tension. Based on some of the ideas presented in Tamiya (1977), Tlusty et al. (1982) developed a computer program simulating chatter of a tandem mill. Variable interstand tension is claimed to be the cause of negative damping and positive damping is due to the variation of contact length on the exit side. By equating the positive damping with the negative damping, a limit of stability that involves various factors such as entry gauge thickness, distance between stands, roll-speed, reduction, and roll-radius was achieved. Further, based on computer simulations of a 3-stand mill, they concluded that above a limiting speed, chatter amplitude increases strongly with roll-speed and that the smaller the friction coefficient at the roll-sheet interface, the larger are the vibrations. In addition the dependence of natural frequency on the roll-speed, friction coefficient, strip width, and strip tension was also studied.

Yarita et al. (1978) examined a 4-DOF linear system modeling the vibrations of a 4-high mill. Rolling force was determined from Hill's equation, and the deformed roll radius was determined by using Hitchcock's formula (see Roberts, 1978). Natural frequencies of the vibrating system were determined numerically. Chatter was found to occur due to degeneration of the lubricant, fluctuations in strip thickness, and entry and exit tensions. A 1-DOF system was also considered with variable stiffness, and a stability criterion was derived that involves damping, mill modulus, and the equivalent spring constant for the sheet.

Misonoh (1980) proposed a 5-DOF mass-spring system that includes the vibrations of the housing in addition to the motions of the four rolls. Numerical results are presented for the natural frequencies for a wide range of housing masses. As the mass of the housing becomes large, the frequency associated with the housing is shown to approach zero. Pawelski et al. (1986) have considered a 5-DOF system and accounted for the variation of damping force due to changing gap geometry through an expanded version of the classical rolling theory by von Karman. The roll-forces were obtained numerically, and nonlinear damping due to plastic deformation of the sheet was found to be important.

Chefneux and Gouzou (1984) considered a 4-DOF model that assumes symmetric chatter in relation to the rolled-strip. Consequently the model reduces to a 2-DOF system. The numerical results indicated that chatter occurs when there is a sudden change in the rolling load due to the presence of welds or a sudden change in lubrication and the sensitivity of the mill to chatter is high. Chatter was concluded to be caused by self-excited vibrations caused by tension in the sheet between the stands.

Nessler and Cory (1989) conducted experimental and finite-element analyses to study fifty-octave backup roll chatter. A planar finite element beam model of the backup rolls, work-rolls and drive spindle was used to study the dynamics of the roll-stack. Based on experimental evidence they conclude that the chatter marks on the backup rolls are the result of a roll stack bending mode being excited during rolling. A relationship was found between frequency of chatter, sheet speeds at which chatter develops, wavelength of the chatter marks, and geometry of the backup rolls. Applications were made to temper mill problems. Guo et al. (1993) also used a 3-D finite-element model to understand the mechanisms and relationships involved in self- and external excitations of the roll-mill system. A simple 4-DOF mass-spring system was also considered as a prelude to the finite-element results. Mill housing was modeled by using 3-D quadrilateral elements, work and backup rolls by 2-D beam elements, and 2-D spring elements to model the interaction between backup roll and work-roll and work-roll and strip. Frequency and modal analysis was conducted, and effects of strip-width, roll-chock excitation, rolling of weld segments, strip modulus, and damping were studied. Further the possibility of strip self-excitation was also considered. Various factors such as excessive entry tension, abrupt changes in rolling, excessive reduction, and small damping have been found to cause third-octave chatter.

More recently Johnson (1994) and Johnson and Qi (1994) approached the problem of chatter using asymptotic methods valid for thin sheets. A rate-dependent material with a relative slip friction condition was considered. An analytical expression for the dynamic contribution to the roll-force due to the vertical vibrations of the work-rolls was found, and the corresponding equivalent damping and spring coefficients associated with the plastic deformation of the sheet were derived. The entry tension was found to affect only the spring coefficient, and the damping coefficient was found to be independent of entry tension. Further, under certain operating conditions, both the damping and spring coefficients were found to become negative. A 2-DOF mass-spring system involving only the vibrations of the work-rolls and a 4-DOF mass-spring system involving the vibrations of both the work and backup rolls were considered, and analytical expressions for the natural frequencies were derived. Further the effect of nonlinear force-displacement relation for the elastic deformation at the interface of the work and backup rolls was discussed.

4.3   PROBLEM FORMULATION

The schematic of a typical 4-high mill stand is shown in Figure 4.1a. During a rolling operation, the rolls can vibrate in either torsional modes or vertical

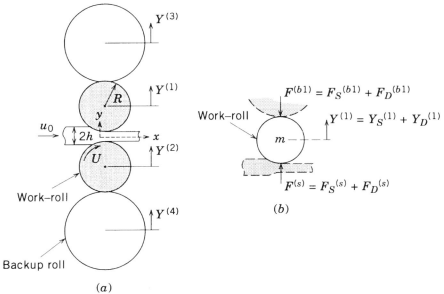

FIGURE 4.1. Schematic of a typical rolling stand: (a) Rolling stand; (b) forces on the upper work-roll.

modes involving either the third-octave chatter or the fifth-octave chatter. In the third octave-chatter, the backup rolls may be involved, whereas in the fifth-octave mode, only the two work-rolls act in unison as the oscillating mass (see Roberts, 1988). In the present work we will confine our attention to only the vertical vibrations of the rolls and not consider the torsional modes. Further the two cases when the backup rolls are fixed and when they are allowed to move vertically along with the work-rolls are considered separately. Due to the complexity of the nonlinear deformation that the two rolls and the strip undergo during rolling, the dynamics of the vibrating system is tractable only when certain simplifications are made with regard to the physical behavior of the system. Accordingly we assume that the vibrations of the entire system can be described by either a 2-DOF model or a 4-DOF model depending on whether the backup rolls are fixed or not. The mass of each of the rolls is lumped at the center of that roll and the interaction between the sheet and work-rolls, and between the work-rolls and the backup rolls, are treated through a system of nonlinear springs and dampers. In this section the governing equations describing the dynamic motion of these lumped masses are presented.

We assume that the vertical vibrations of the rolls are small and that their order of magnitude is represented by an infinitesimal nondimensional parameter ε. Let $\hat{Y}^{(1)}$ and $\hat{Y}^{(2)}$ denote the vertical displacements of the top and bottom work rolls, respectively, and $\hat{Y}^{(3)}$ and $\hat{Y}^{(4)}$ denote the vertical displacements of the top and bottom backup rolls, respectively, with all the displacements being measured from the rest position. We assume that these displacements can be separated into steady and dynamic parts, namely $\hat{Y}^{(1)} = \hat{Y}_S + \varepsilon\hat{Y}_D^{(1)}$, $\hat{Y}^{(2)} = -\hat{Y}_S + \varepsilon\hat{Y}_D^{(2)}$, $\hat{Y}^{(3)} = \hat{Y}_S + \varepsilon\hat{Y}_D^{(3)}$, and $\hat{Y}^{(4)} = -\hat{Y}_S + \varepsilon\hat{Y}_D^{(4)}$. The static parts are the displacements of the rolls during steady rolling of the sheet and the dynamic parts are associated with the chatter of the rolls. Due to chatter, the gap between the work rolls is time-dependent and is given by $2\hat{H}(\hat{x}, \hat{t}) = 2\hat{h}(\hat{x}) + \varepsilon[\hat{Y}_D^{(1)}(\hat{t}) - \hat{Y}_D^{(2)}(\hat{t})]$, where $2\hat{h}(\hat{x})$ is the gap between the rolls under steady state rolling and $2\hat{b}(t) = \varepsilon[\hat{Y}_D^{(1)}(\hat{t}) - \hat{Y}_D^{(2)}(\hat{t})]$ is the change in thickness of the gap due to chatter.

The net forces acting on the rolls also consist of steady and dynamic parts, as shown in Figure 4.1b. The upper work roll is subjected to a force from the sheet given by $\hat{F}^{(s)} = \hat{F}_S + \hat{F}_D^{(s)}$, and a force from the backup roll given by $\hat{F}^{(b_1)} = \hat{F}_S + \hat{F}_D^{(b_1)}$. Similar forces act on the lower work roll and are denoted with a superscript b_2. The roll-stand housing and foundation exert forces (not shown in Figure 4.1b) on the upper backup rolls and is given by $\hat{F}^{(h_3)} = \hat{F}_S + \hat{F}_D^{(h_3)}$. A similar force acts on the lower backup roll and is denoted with a superscript h_4. Since the problem is two-dimensional, we note that all of the forces are per unit length of the rolls.

For a 2-DOF system, the backup rolls are assumed to be rigidly supported at their centers by the roll stand housing and only the work-rolls are free to vibrate vertically. Therefore in this case $Y^{(3)}$ and $Y^{(4)}$ are zero, and the governing equations of motion for the two work-rolls are given by

$$\varepsilon \hat{m} \frac{d^2 \hat{Y}_D^{(1)}}{d\hat{t}^2} = \hat{F}_D^{(s)} - \hat{F}_D^{(b_1)}, \tag{1}$$

$$\varepsilon \hat{m} \frac{d^2 \hat{Y}_D^{(2)}}{d\hat{t}^2} = -\hat{F}_D^{(s)} + \hat{F}_D^{(b_2)}. \tag{2}$$

In the above, $\hat{m}$ is the mass per unit length of the work-rolls.

For a 4-DOF system the work and backup rolls are both free to vibrate in the housing. The governing equations in this case are given by

$$\varepsilon \hat{m} \frac{d^2 \hat{Y}_D^{(1)}}{d\hat{t}^2} = \hat{F}_D^{(s)} - \hat{F}_D^{(b_1)}, \tag{3}$$

$$\varepsilon \hat{m} \frac{d^2 \hat{Y}_D^2}{d\hat{t}^2} = -\hat{F}_D^{(s)} + \hat{F}_D^{(b_2)}, \tag{4}$$

$$\varepsilon \hat{m}^{(b)} \frac{d^2 \hat{Y}_D^{(3)}}{d\hat{t}^2} = \hat{F}_{(D)}^{(s)} + \hat{F}_{(D)}^{(b_1)}, \tag{5}$$

$$\varepsilon \hat{m}^{(b)} \frac{d^2 \hat{Y}_D^{(4)}}{d\hat{t}^2} = \hat{F}_{(D)}^{(h_4)} - \hat{F}_{(D)}^{(b_2)}, \tag{6}$$

where $\hat{m}^{(b)}$ is the mass per unit length of the backup rolls.

The force $\hat{F}^{(s)}$ that the sheet exerts on the work-rolls is obtained by considering the inelastic deformation of the sheet as it passes through the roll-bite gap. The steady state part of this force, $\hat{F}_S^{(s)}$, is well-known, and it is the dynamic contribution $\hat{F}_D^{(s)}$ that is needed to solve the above equations. In a later section closed-form analytic expressions for $\hat{F}^{(s)}$ and hence $\hat{F}_D^{(s)}$ are obtained by means of a regular perturbation method.

4.4   CALCULATION OF BACKUP ROLL-FORCE

We begin by considering the 2-DOF system and use the popular lumped-mass technique where the complex dynamic motion of an elastic structure is described by placing all the inertial properties in a lumped mass and deformation is treated through an equivalent elastic spring and damper system. To calculate the backup roll-force, we assume that the deformation between the work- and backup rolls is elastic. The relative displacement between the centers of two rolls of the same material due to a compressive load $\hat{F}^{(b_i)}$ ($i = 1, 2$) per unit length is well known and given in Roark and Young (1975)

$$\hat{Y}^{(i)} = 2\hat{F}^{(b_i)} \frac{1 - v^2}{\hat{E}} \left[\frac{2}{3} + \ln \left\{ \frac{0.78 \hat{E} (\hat{D}_1 + \hat{D}_2)}{(1 - v^2) \hat{F}^{(b_i)}} \right\} \right], \qquad i = 1, 2, \tag{7}$$

where $\hat{D}_1$ and $\hat{D}_2$ are the work-roll and backup roll diameters, respectively (we assume that the top and bottom rolls have the same diameters), E is Young's modulus, and v is Poisson's ratio of the rolls. We note that the relation between $Y^{(i)}$ and $F^{(b_i)}$ given by (7) is nonlinear and typical of contact problems. Equation (7) indicates that it is convenient to nondimensionalize $\hat{Y}^{(i)}$ by $2\hat{F}_S(1 - v^2)E^{-1}$. Further, after nondimensionalizing the forces by the steady state force $\hat{F}_S$, Equation (7) can be rewritten as

$$F^{(b_i)} = \frac{\hat{F}^{(b_i)}}{\hat{F}_S} = 1 + \frac{\hat{F}_D^{(b_i)}}{\hat{F}_S} = 1 + F_D^{(b_i)}, \tag{8}$$

$$Y^{(i)} = \pm(1 + F_D^{(b_i)})\{Y_S - \ln(1 + F_D^{(b_i)})\}, \tag{9}$$

where the steady displacement due to $\hat{F}_S$ is

$$Y_S = \frac{2}{3} + \ln\left\{\frac{0.78\hat{E}(\hat{D}_1 + \hat{D}_2)}{(1 - v^2)\hat{F}_S}\right\}. \tag{10}$$

Recalling that $Y^{(i)} = Y_S^{(i)} + \varepsilon Y_D^{(i)}$, we can write the dynamic parts of the displacement as

$$\varepsilon Y_D^{(i)} = \pm Y_S F_D^{(b_i)} \mp (1 + F_D^{(b_i)}) \ln(1 + F_D^{(b_i)}). \tag{11}$$

The above two equations for $Y_D^{(i)}$ coupled with the nondimensionalized version of Equations (1) and (2), which are expressed as

$$\varepsilon\frac{d^2 Y_D^{(i)}}{dt^2} = \pm F_D^{(s)} \mp F_D^{(b_i)}, \tag{12}$$

give four equations in the four unknowns $Y_D^{(i)}$ and $F_D^{(b_i)}$ with i taking on values 1 and 2 for the top and the bottom backup rolls, respectively.

By taking the average and difference of Equation (12), equations describing the mass-center motion $\bar{Y}(t) = \frac{1}{2}(Y_D^{(1)} + Y_D^{(2)})$ and the squeezing motion of the roll-bite gap $B(t) = \frac{1}{2}(Y_D^{(1)} - Y_D^{(2)})$ are obtained as

$$\varepsilon\frac{d^2 \bar{Y}}{dt^2} = \frac{1}{2}(F_D^{(b_2)} - F_D^{(b_1)}) \quad\text{and}\quad \varepsilon\frac{d^2 B}{dt^2} = F_D^s - \frac{1}{2}(F_D^{(b_2)} + F_D^{(b_1)}). \tag{13}$$

In obtaining the above equations, time is nondimensionalized by $(2\hat{m}(1 - v^2)/(\pi\hat{E}))^{1/2}$. Further it is evident from these equations that the mass-center motion is not affected by the dynamic sheet force. The mass-center motion is associated with the corrugations in the sheet, and the squeezing motion causes gauge variations.

The steady state roll-force $\hat{F}_S$ exerted by the sheet on the rolls (and hence Y_S given by Equation (10) and occurring in Equation (11)) and the corresponding dynamic force $F_D^{(s)}$ due to chatter (occurring in Equation (12)) are obtained by considering the sheet motion through the roll-bite. We record here the expression for these quantities and defer their derivation to a later section (Section 4.7):

$$\hat{F}_S = \hat{\tau}_0 \delta^{-1} \hat{l} \left[\int_0^1 N_S \, dx - \sigma_S^{entry} \right], \tag{14}$$

$$F_D^{(s)} = -\varepsilon [K^{(s)} \eta B + S^{(s)} \eta \alpha \dot{B} + \phi \sigma_D^{entry}(t)], \tag{15}$$

where N_S is essentially the steady part of roll pressure and given by (57), σ_S^{entry} and $\sigma_D^{entry}(t)$ are the steady and dynamic parts of the applied entry tension respectively, δ (thinness parameter) is the ratio of entry gauge $\hat{h}_0$ of the sheet and the length of roll-bite $\hat{l}$. Further $K^{(s)}$ is the effective spring coefficient due to the sheet deformation and is given by Equation (66), and $S^{(s)}$ is the effective damping coefficient given by Equation (67). The constants η, α, and ϕ are defined as

$$\eta = \frac{2(1-v^2)\hat{\tau}_0 \hat{l}}{\pi \hat{E} \hat{h}_0 \delta}, \quad \alpha = \frac{\hat{l}}{\hat{u}_0} \sqrt{\frac{\pi \hat{E}}{2\hat{m}(1-v^2)}}, \quad \text{and} \quad \phi = \frac{\hat{\tau}_0 \hat{l}}{\delta \hat{F}_S}. \tag{16}$$

Values of the various parameters for a typical rolling process are given in Table 4.1. For these values we find $\eta = O(10^{-2})$ and $\alpha = O(10^2)$. In addition, since $\delta^{-1} \hat{\tau}_0 \hat{l}$ is the characteristic force per unit length, $\phi = O(1)$.

Equations (11) are nonlinear and therefore explicit solutions for $Y_D^i(t)$ are difficult to find. However, when $F_D^{(b_i)}(t)$ are small compared to unity, linearization is possible. Next we examine the linear approximation for both 2-DOF and 4-DOF models, and following that the effect of nonlinearity is discussed.

4.5 LINEAR THEORY

4.5.1 2-DOF System

When the dynamic forces $F_D^{(b_i)}(t)$ that the backup rolls exert on the work-rolls are small compared to unity, Equations (11) can be approximated as

$$\varepsilon Y_D^{(i)} = \pm (Y_S - 1) F_D^{(b_i)}(t).$$

Substituting these in (13), we get the following linear equations of motion:

$$\frac{d^2 \bar{Y}}{dt^2} + \frac{\bar{Y}}{Y_S - 1} = 0, \tag{17}$$

$$\frac{d^2 B}{dt^2} + S^{(s)} \eta \alpha \frac{dB}{dt} + \left[\frac{1}{Y_S - 1} + K^{(s)} \eta \right] B = -\phi \sigma_D^{entry}(t), \tag{18}$$

for the mass-center motion and the squeezing motion of the roll-bite gap, respectively. In obtaining the second equation above, we used Equation (15) for $F_D^{(s)}$. Equations (17) and (18) indicate that the mass-center motion is oscillatory (as long as $Y_S > 1$) and undamped. Further the net spring coefficient for the mass-center motion is $(Y_S - 1)^{-1}$ which is independent of the dynamic part of the sheet deformation in the roll-bite. The squeezing motion is, however, a damped motion, with the damping coefficient originating from the dynamic part of the sheet deformation alone. The net spring coefficient has contributions from both the elastic deformation at the interface between the work and backup rolls and the sheet deformation. In Figure 4.2, the effective

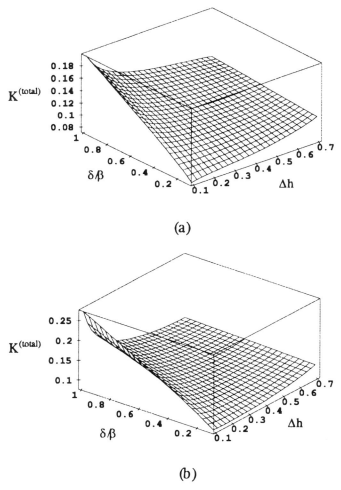

(a)

(b)

FIGURE 4.2. Total effective spring-coefficient $K^{(tot)}$ versus reduction and δ/β: (a) $n = 4$; (b) $n = 100$.

coefficient for the squeezing motion, $K^{(tot)} = (Y_S - 1)^{-1} + K^{(s)}\eta$ is plotted against reduction and δ/β. Comparing this with Figure 4.8 where $K^{(s)}$ is plotted against reduction and δ/β, we note that even though $K^{(s)}$ is negative for small δ/β and small reductions, the total effective spring coefficient $K^{(tot)}$ is positive for all reductions and δ/β values. However, it is clear from Figure 4.9, where $S^{(s)}$ is plotted against reduction and δ/β, that the damping coefficient can be negative when δ/β is small (small friction) or when n is large (weak rate dependence).

From the linear theory of vibration, the nondimensional natural frequency of the mass-center motion (Equation (17)) is given by

$$\omega_0 = \sqrt{\frac{1}{Y_S - 1}}. \tag{19}$$

Equation (18) has a general solution of the form

$$B(t) = e^{-st}(Ae^{\sqrt{s^2 - \omega_n^2}\,t} + Be^{-\sqrt{s^2 - \omega_n^2}\,t}), \tag{20}$$

where

$$s = \frac{\eta\alpha S^{(s)}}{2} \quad \text{and} \quad \omega_n = \sqrt{\frac{1}{Y_S - 1} + K^{(s)}\eta}. \tag{21}$$

Note that ω_n is the natural frequency of the squeezing motion when damping is negligible. Following well-known linear vibration theory, several observation can be made from Equation (20). First, if s is negative, we have an exponentially growing amplitude; the vibrations become unstable and chatter results. As evident from Figure 4.9, $S^{(s)}$ (and hence s) is negative for weakly rate-sensitive materials or when the reduction and friction are large. A similar situation is possible when $K^{(s)}$ is sufficiently negative to make the natural frequency ω_n imaginary. However, for practical situations this may not be possible unless the entry tension is sufficiently large to decrease $K^{(s)}$ (Equation (66)). When both s and ω_n are positive, overdamping is possible when $s > \omega_n$, and underdamping is possible when $s < \omega_n$, with critical damping occurring when $s = \omega_n$. In all these cases, however, the amplitude is exponentially decreasing, and hence no self-sustained vibrational motion results. For the values shown in Table 4.1, the characteristic time (i.e., the time scale associated with the elastic restoring force and also the quantity with which time was nondimensionalized) is of the order 10^{-4} s, indicating that when damping is positive, the vibrations are damped very quickly, and that when damping is negative, the vibrations increase with time rapidly until the damping becomes positive again.

Further, beating phenomena may be possible when damping is negligible. From Equation (20) we note that in the absence of damping, the frequency of

TABLE 4.1
Vaues of Various Parameters in a Typical Rolling Process

$\hat{E}$	$3 \times 10^7 \, \text{psi}$
$\hat{m}$	0.33 slug/inch
$\hat{u}_0$	100 inches
$\hat{l}_0$	1 inch
$\hat{h}_0$	0.1 inch
$\hat{D}_1$	24 inches
$\hat{D}_2$	60 inches
$\hat{p}_0$	$5 \times 10^4 \, \text{psi}$
$\hat{F}_s$	$5 \times 10^4 \, \text{lb/inch}$

the squeezing motion is ω_n, which can be written as

$$\omega_n = \omega_0 \sqrt{1 + K^{(s)} \eta (Y_S - 1)}.$$

This expression implies that beating is possible with a frequency of $(\omega_n + \omega_0)/2$ and that an amplitude modulation of $(\omega_n - \omega_0)/2$ when $K^{(s)}\eta(Y_S - 1)$ is small compared to unity. For the data in Table 4.1, we note that $\eta = O(10^{-2})$, $Y_S \cong 11.3$, and from Figure 4.8, $K^{(s)}$ has a typical value of 2 when the reduction is 20–40%. Thus we note that $\omega_n \cong \omega_0$, implying that beating phenomenon is possible if damping is negligible. In general, ω_n can vary anywhere from approximately ω_0 to $1.5\omega_0$ since $K^{(s)}$ ranges from 0 to 20 for the data shown in Table 4.1.

For the values shown in Table 4.1, $Y_S \cong 11.3$, and the frequency associated with the mass-center motion can be found to be $\hat{\omega}_0 \cong 620 \, \text{Hz}$, which falls in the range of the fifth-octave mode. This is consistent with the roll-mill observations where the fifth-octave chatter mode has been associated with the work-roll vibrations. The corresponding frequency ω_n associated with the squeezing motion can range anywhere from $620 \, \text{Hz}$ to approximately $900 \, \text{Hz}$. Thus the squeezing mode motion may or may not fall in the fifth-octave mode chatter depending on the level of friction and reduction.

It is also worth noting that for the values shown in Table 4.1, the characteristic time (i.e., the time scale associated with the elastic restoring force and also the quantity with which the dimensional time was nondimensionalized) is of the order 10^{-4} s, indicating that when damping is positive (e.g., when rate-dependence is strong such as in hot-rolling processes), the vibrations are damped out very quickly, and that when damping is negative (e.g., when rate-dependence is weak such as in cold-rolling processes), the vibrations increase with time rapidly until the damping becomes positive again. Thus, since all of the damping comes from the plastic deformation of the sheet in the roll-bite gap, we conclude that the sheet deformation is an important factor that cannot be ignored in chatter analyses. This is in contrast to several of the previous studies on chatter, which do not take into account the plastic deformation of the sheet in the roll-bite.

4.5.2 4-DOF System

The equivalent mass-spring system for a 4-high rolling stand with 4-DOF is shown in Figure 4.3. Again we assume that the forces that the backup rolls exert on the work rolls $F_D^{(b_i)}(t)$ are small compared to unity. In this case, noting that the distance between the centers of the work- and backup rolls is given by $Y^{(1)} - Y^{(3)}$ for the upper rolls and $Y^{(2)} - Y^{(4)}$ for lower rolls, the linear expressions for $F_D^{(b_i)}(t)$ are given by

$$Y_D^{(1)} - Y_D^{(3)} \cong (Y_S - 1)F_D^{(b_i)}(t) \quad \text{and} \quad Y_D^{(2)} - Y_D^{(4)} \cong -(Y_S - 1)F_D^{(b_2)}(t). \quad (22)$$

Since one interest is to determine the natural frequencies of the system, we neglect damping and also assume that the entry tension is zero. In this case the dynamic sheet force is given by $\frac{1}{2}K^{(s)}\eta(Y_D^{(1)} - Y_D^{(2)})$, and the dynamic force from the housing is given by $K^{(h)}Y_D^{(3)}$ for the upper backup roll and $K^{(h)}Y_D^{(4)}$ for the bottom. Here $K^{(h)}$ is the effective spring coefficient due to the housing. The

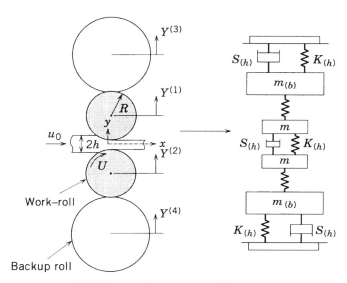

FIGURE 4.3. Four-degree-of-freedom system.

governing equations then become

$$\frac{d^2 Y_D^{(1)}}{dt^2} + \omega_0^2(Y_D^{(1)} - Y_D^{(3)}) + \frac{1}{2}K^{(s)}\eta(Y_D^{(1)} - Y_D^{(2)}) = 0,$$

$$\frac{d^2 Y_D^{(2)}}{dt^2} + \omega_0^2(Y_D^{(2)} - Y_D^{(4)}) + \frac{1}{2}K^{(s)}\eta(Y_D^{(2)} - Y_D^{(1)}) = 0,$$

$$\frac{\hat{m}^{(b)}}{\hat{m}}\frac{d^2 Y_D^{(3)}}{dt^2} + K^{(h)}Y_D^{(3)} + \omega_0^2(Y_D^{(3)} - Y_D^{(1)}) = 0,$$

$$\frac{\hat{m}^{(b)}}{\hat{m}}\frac{d^2 Y_D^{(4)}}{dt^2} + K^{(h)}Y_D^{(4)} + \omega_0^2(Y_D^{(4)} - Y_D^{(2)}) = 0,$$

where $\omega_0^2 = (Y_S - 1)^{-1}$. Substituting $Y_D^{(i)} = A_i e^{i\omega t}$ into the above and denoting $\lambda = \hat{m}^{(b)}/\hat{m}$, the characteristic equation determining the natural frequencies is given by

$$\det\begin{bmatrix} K^{(h)} + \omega_0^2 - \lambda\omega^2 & -\omega_0^2 & 0 & 0 \\ -\omega_0^2 & \frac{K^{(s)}\eta}{2} + \omega_0^2 - \omega^2 & -\frac{K^{(s)}\eta}{2} & 0 \\ 0 & -\frac{K^{(s)}\eta}{2} & \frac{K^{(s)}\eta}{2} + \omega_0^2 - \omega^2 & -\omega_0^2 \\ 0 & 0 & -\omega_0^2 & K^{(h)} + \omega_0^2 - \lambda\omega^2 \end{bmatrix} = 0.$$

From this, the natural frequencies are obtained as

$$\omega^{(1,2)} = \frac{1}{\sqrt{2}}\left\{\left(\omega_n^2 + \frac{1}{\lambda}\omega_h^2\right) \pm \sqrt{\left(\omega_n^2 + \frac{1}{\lambda}\omega_h^2\right)^2 - \frac{4}{\lambda}(\omega_0^2 K^{(s)}\eta + K^{(h)}\omega_n^2)}\right\}^{1/2}$$

$$\omega^{(3,4)} = \frac{1}{\sqrt{2}}\left\{\left(\omega_0^2 + \frac{1}{\lambda}\omega_h^2\right) \pm \sqrt{\left(\omega_0^2 + \frac{1}{\lambda}\omega_h^2\right)^2 - \frac{4}{\lambda}\omega_0^2 K^{(h)}}\right\}^{1/2}$$

In the above, $\omega_n^2 = \omega_0^2 + K^{(s)}\eta$ and $\omega_h^2 = \omega_0^2 + K^{(h)}$. It is worth noting that the above frequencies reduce to those for the 2-DOF model when λ is large, namely when the mass of the backup rolls is much bigger than that of the work-rolls. The corresponding eigen-modes are given by

$$Q_k = \left\{1, \frac{B_{k_1}}{\omega_0^2}, \frac{B_{k_1}^2 K^{(s)}\eta}{2\omega_0^2(B_{k_1}B_{k_2} - \omega_0^4)}, \frac{B_{k_1}^2 K^{(s)}\eta}{2(B_{k_1}B_{k_2} - \omega_0^4)}\right\}^T, \quad k = 1.4,$$

with

$$B_{k_1} = \omega_h^2 - \lambda\omega^{(k)2} \quad \text{and} \quad B_{k_2} = \tfrac{1}{2}K^{(s)}\eta + \omega_0^2 - \omega^{(k)2}.$$

The four modes are illustrated in Figure 4.4. The first two modes with the

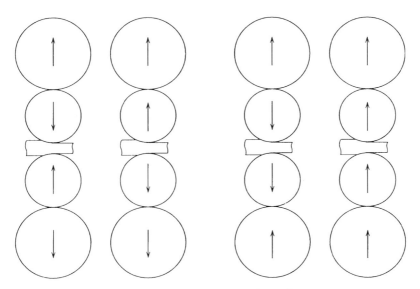

FIGURE 4.4. Eigen-mode motions.

frequencies $\omega^{(1,2)}$ are squeezing motions. In the first mode, the backup rolls move in the opposite direction to their neighboring work-rolls. For the second mode, each backup roll moves in the same direction as the neighboring work-roll. In the third and fourth modes with the frequencies $\omega^{(3,4)}$, the work-rolls move in the same direction, a mass-center motion. In the fourth mode, all four rolls move in the same direction, whereas in the third mode, each backup roll moves in a direction opposite to that of the neighboring work-roll. For the data given in Qi and Johnson (1994), the natural frequencies can be found to be $\hat{\omega}^{(1)} = 633\,\text{Hz}$, $\hat{\omega}^{(2)} = 157\,\text{Hz}$, $\hat{\omega}^{(3)} = 309\,\text{Hz}$, and $\hat{\omega}^{(4)} = 110\,\text{Hz}$. The first mode lies in the range of fifth-octave chatter, and the work-rolls vibrate in a squeezing fashion with a much larger amplitude than the backup rolls. The second, third, and fourth modes, however, lie in the range of third-octave chatter, with the third and fourth contributing toward mass-center motion. For further details the reader is referred to Qi and Johnson (1994).

4.6 NONLINEAR EFFECTS

By eliminating the displacements from Equations (11) and (12), we get equations involving only the forces:

$$[Y_S - 1 - \ln(1 + F_D^{b_i}(t))]\ddot{F}_D^{b_i}(t) - \frac{(\dot{F}_D^{b_i}(t))^2}{1 + F_D^{b_i}(t)} + F_D^{b_i}(t) = F_D^s, \qquad (23)$$

where $i = 1, 2$.

The nonlinear effect associated with the deformation at the work-backup roll interface and given by Equation (23) was considered in detail by Qi and Johnson (1994). Here a brief review of this work is presented. To facilitate analysis, the dynamic sheet force $F_D^{(s)}$ is assumed to be negligible. In this case there is no coupling between the rolls, and the top and bottom work-rolls vibrate independent of each other. Consequently the governing equations become

$$[Y_S - 1 - \ln(1 + F_D^{(b_i)}(t))]\ddot{F}_D^{(b_i)}(t) - \frac{(\dot{F}_D^{(b_i)}(t))^2}{1 + F_D^{(b_i)}(t)} + F_D^{(b_i)}(t) \cong 0, \qquad (24)$$

where $i = 1, 2$. In practice (evident from the previous section), $\mu := \omega_0^2 = (Y_S - 1)^{-1}$ is a small quantity, and we examine the solution behavior of the above equation for small μ. By introducing a rescaled time $\tau = \mu^{1/2}t$ into Equation (24), we obtain

$$[1 - \mu \ln(1 + F_D^{(b_i)}(\tau))]\ddot{F}_D^{(b_i)}(\tau) - \mu \frac{(\dot{F}_D^{(b_i)}(\tau))^2}{1 + F_D^{(b_i)}(\tau)} + F_D^{(b_i)}(\tau) \cong 0, \qquad (25)$$

where $i = 1, 2$. The method of Poincare-Lindstedt, which takes into account the cumulative effect of the weak nonlinearity, is used to obtain an asymptotic solution to the above equations. Solutions are sought in the form

$$F^{(b_1)} = F_0(\tilde{\tau}) + \mu F_1(\tilde{\tau}) + \cdots$$

and

$$\tilde{\tau} = (1 + \mu\omega_1 + \mu^2\omega_2 + \cdots)\tau.$$

The parameters $\omega_i(i = 1, 2, \ldots)$ are the frequency shift parameters and are chosen such that the secular terms arising after substituting the above expansions into (25) are suppressed. Following the procedure in Qi and Johnson (1994), the first frequency shift parameter ω_1 can be shown to be given by

$$\omega_1 = -\frac{1}{2}\left[\frac{1}{2} - \frac{1}{a_0^2} - \ln(1 + a_0) + k(a_0)\right],$$

where a_0 is the initial value of $F_D^{(b_1)}$ and

$$k(a_0) = \frac{1}{2\pi a_0^2} \int_0^{2\pi} \frac{1 - a_0^2 - a_0^3\tau \sin\tau}{1 + a_0 \cos\tau} d\tau.$$

The frequency parameter ω_1 is plotted in Figure 4.5. Since the actual frequency shift is $\mu\omega_1$, we note from the figure that the nonlinearity associated with the deformation at the work–backup roll interface decreases the frequency. How-

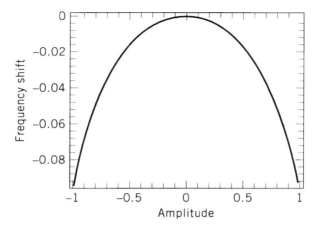

FIGURE 4.5. Frequency shift ω_1 against amplitude a_0.

ever, since μ is typically small, this decrease is only slight. Consequently the primary effect of the nonlinearity is to trigger higher harmonic modes of vibration. The first few terms are given by

$$F_1(t) = a_1 \cos \tilde{\tau} + a_2 \cos 2\tilde{\tau} + a_3 \cos 3\tilde{\tau} + \cdots,$$

where

$$a_1 = -\frac{4}{3} - \frac{3}{4a_0^3} + \frac{8}{9a_0^2} + \frac{9}{8a_0} - \frac{9a_0}{32} + \frac{(1 + a_0)^2}{36a_0^3} \sqrt{\frac{1 - a_0}{1 \ a_0}} \, (27 - 59a_0 + 32a_0^2),$$

$$a_2 = \frac{4}{3} - \frac{8}{9a_0^2} + \frac{8(1 - a_0^2)^{3/2}}{9a_0^2} \quad \text{and} \quad a_3 = \frac{3}{4a_0^3} - \frac{9}{8a_0} + \frac{9a_0}{32} - \frac{3(1 - a_0^2)^{3/2}}{4a_0^3}.$$

The amplitudes a_i are plotted against initial amplitude a_0 in Figure 4.6. Again, since a_1 is negative, the nonlinearity tends to decrease the amplitude of the primary mode of vibration.

4.7   CALCULATION OF SHEET FORCE

In this section analytical expressions for the spring and damping coefficients $K^{(s)}$ and $S^{(s)}$ due to the sheet deformation are derived using a regular perturbation expansion. The application of asymptotic methods requires that a set of dimensionless variables be defined. The natural length scales in x and y directions for the sheet-deformation problem are the entry gauge $\hat{h}_0$ and the roll-bite gap length $\hat{l}$. Therefore the nondimensionalization that we follow in this section is different from the one introduced previously for the vibrational motion of the rolls.

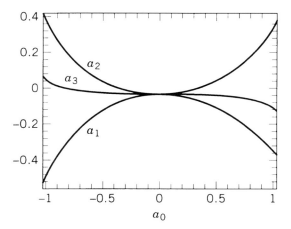

FIGURE 4.6. Amplitude of higher-order terms a_1, a_2, and a_3 plotted against a_0.

During steady state rolling the material enters the roll-bite at $\hat{x} = 0$ at a velocity $\hat{u}_0$ and exits at $\hat{x} = \hat{l}$. The thickness of the material is $2\hat{h}_0$ at the entry point and $2\hat{h}_1$ at the exit point, with the difference $2\hat{h}_0 - 2\hat{h}_1 = 2\Delta\hat{h}$ being the total reduction. The steady state gap between the rolls is described by $\hat{y} = \pm\hat{h}(x)$, and $\hat{U}$ is the roll-speed. During chatter, however, the gap between the rolls is given by $2\hat{H}(\hat{x}, \hat{t}) = 2\hat{h}(\hat{x}) + \varepsilon[\hat{Y}_D^{(1)}(\hat{t}) - \hat{Y}_D^{(2)}(\hat{t})]$, and the entry point is time dependent (see Figure 4.7) while the exit point remains fixed at $\hat{x} = \hat{l}$ where the gap is a minimum. Since, in practice, the sheet has a self-centering tendency, it is reasonable to assume that the sheet centerline remains coincident with the centerline of the gap throughout the chatter motion. Consequently the sheet first contacts the upper and lower rolls at the same position $\hat{x} = \hat{a}(\hat{t})$, with the length of the inelastically deforming region in the roll-bite being $\hat{a}(\hat{t}) \leqslant \hat{x} \leqslant \hat{l}$.

We assume that the elastic and inertial effects can be ignored and that a rate-dependent power law governs the inelastic behavior of the sheet material. Then the nondimensionsl governing equations for plane-strain deformation are the equilibrium equations

$$-\frac{\partial p}{\partial x} + \frac{\partial S_{xy}}{\partial y} + \frac{\delta}{\beta}\frac{\partial S_{xx}}{\partial x} = 0 \qquad (26)$$

$$-\frac{\partial p}{\partial y} + \delta^2\frac{\partial S_{xy}}{\partial x} + \frac{\delta}{\beta}\frac{\partial S_{yy}}{\partial y} = 0, \qquad (27)$$

the conservation of mass equation

$$\delta\frac{\partial u}{\partial x} + \frac{\partial v}{\partial y} = 0, \qquad (28)$$

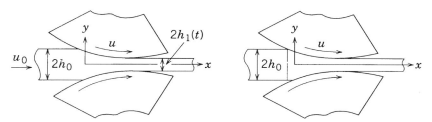

FIGURE 4.7. Squeezing motion of the roll-bite during chatter.

and the rate-dependent constitutive equations

$$\frac{\partial u}{\partial x} = \Omega^{n-1} S_{xx}, \tag{29}$$

$$\frac{\partial v}{\partial y} = \delta \Omega^{n-1} S_{yy}, \tag{30}$$

$$\frac{\partial u}{\partial y} + \delta \frac{\partial v}{\partial x} = 2\beta \delta \Omega^{n-1} S_{xy}, \tag{31}$$

where

$$\Omega = \left[\tfrac{1}{2}(S_{xx}^2 + S_{yy}^2) + \beta^2 S_{xy}^2\right]^{1/2}, \tag{32}$$

$$\beta = \frac{\hat{\tau}_0}{\hat{S}_0} \quad \text{and} \quad \delta = \frac{\hat{h}_0}{\hat{l}}. \tag{33}$$

In obtaining the above equations, the $\hat{x}$ and $\hat{y}$ coordinates are nondimensionalized by $\hat{l}$ and $\hat{h}_0$, respectively. The deviatoric stresses $\hat{S}_{xx}$ and $\hat{S}_{yy}$ are nondimensionalized by a characteristic longitudinal deviatoric stress $\hat{S}_0$, the shear stress $\hat{S}_{xy}$ by a characteristic shear stress $\hat{\tau}_0$, and the pressure by $\delta^{-1}\hat{\tau}_0$. In addition the velocity components $\hat{u}$ and $\hat{v}$ are nondimensionalized by the entry speed of the material $\hat{u}_0$. The factor n appearing in the constitutive equations above is the rate-sensitivity parameter. The dimensionless constitutive equations are obtained from the corresponding dimensional constitutive equation, $\hat{D}_{ij} = \hat{B}\hat{\Omega}^{n-1}\hat{S}_{ij}$ by defining the characteristic longitudinal deviatoric stress $\hat{S}_0 = (\hat{u}_0/\hat{B}\hat{l})^{1/n}$. Here $\hat{B}$ is a material parameter, $\hat{D}_{ij}$ are the components of the deformation-rate tensor, and $\hat{\Omega} = \sqrt{\tfrac{1}{2}\hat{S}_{ij}\hat{S}_{ij}}$. For future reference, we will also record here that $2\hat{b}(\hat{t}) = \hat{Y}_D^{(1)}(\hat{t}) - \hat{Y}_D^{(2)}(\hat{t})$, $\hat{Y}_1$ and $\hat{Y}_2$ are nondimensionalized by $\hat{h}_0$, $\hat{a}(\hat{t})$ by $\hat{l}$, and time $\hat{t}$ by $\hat{l}/\hat{u}_0$.

The boundary conditions are as follows: At the entry and exit points, the longitudinal stress component must balance the applied tensions, i.e.

$$\int_{H^-(a(t),t)}^{H^+(a(t),t)} \sigma_{xx}(a(t), \; y, \; t)dy = 2\sigma^{\text{entry}}(t),$$

$$\int_{H^-(,t)}^{H^+(1,t)} \sigma_{xx}(1, \; y, \; t)dy = 2\sigma^{\text{exit}}(t)h_1. \tag{34}$$

where $H^+(x, t) = h(x) + \varepsilon Y_1(t)$, $H^-(x, t) = -h(x) + \varepsilon Y_2(t)$. At the two interfaces $y = H^\pm(x, t)$, the friction condition is applied. Here we assume a relative-slip model for the friction at the interfaces between the rolls and the sheet material. In dimensional terms this can be expressed as $\hat{\tau}_s = \hat{\kappa}(\hat{U} - \hat{u}_s)$ for the upper interface where $\hat{\tau}_s$ is the tangential component of the traction at the roll-sheet interface, $\hat{u}_s$ is the tangential speed of the sheet at the interface, and $\hat{\kappa}$ is a coefficient of friction. A similar relation holds for the lower interface between the sheet and the roll. The traction $\hat{\tau}_s$ can be shown to be approximately equal to $\hat{S}_{xy}$ (see Equation (39) and Cherukuri and Johnson, 1996, for details). Hence we note that the characteristic shear stress $\hat{\tau}_0$ is determined by the friction condition and can be taken as $\hat{\kappa}\Delta\hat{U}$, where $\Delta\hat{U}$ is a measure of the relative velocity between the rolls and the sheet. Thus the boundary conditions at the interface are given by

$$\tau_s = \frac{U - u_s}{\Delta U} \text{ at } y = H^+(x, t) \quad \text{and} \quad \tau_s = -\frac{U - u_s}{\Delta U} \text{ at } y = H^-(x, t), \tag{35}$$

where $\Delta U = \Delta\hat{U}/\hat{u}_0$. In addition we have the kinematic boundary condition or no penetration condition at the interfaces as

$$\mathbf{u} \cdot \mathbf{n} = \varepsilon \dot{Y}_1(t)\mathbf{e}_2 \cdot \mathbf{n} \text{ at } y = H^+(x, t) \quad \text{and} \quad \mathbf{u} \cdot \mathbf{n} = \varepsilon \dot{Y}_2\mathbf{e}_2 \cdot \mathbf{n} \text{ at } y = H^-(x, t), \tag{36}$$

where u is the velocity vector, $\mathbf{n}$ is the unit normal to the roll-surface, and $\mathbf{e}_2$ is a unit vector in y-direction. By substituting appropriate expressions for $\mathbf{n}$ in the above, it is trivial to show that Equations (36) can also be expressed as

$$v - \delta h'u = \delta\varepsilon \dot{Y}_1(t) \text{ at } y = H^+(x, t) \quad \text{and} \quad v + \delta h'u = \delta\varepsilon \dot{Y}_2(t) \text{ at } y = H^-(x, t). \tag{37}$$

The force that the sheet exerts on the work-rolls is calculated by solving the Equations (26)–(31) subject to the boundary conditions (34), (35), and (37). This is done in detail next.

In the following we assume that the thinness parameter $\delta \ll 1$ and that the friction is weak so that $\beta \ll 1$. The first assumption is based on the observation that in a typical rolling process, δ is often of the order of 0.1–0.3. The weak friction assumption is made to keep the mathematics tractable and make analytical solutions possible. Further we assume that β is not too small so that $\delta/\beta \leqslant O(1)$. Moderate to large friction cases are highly nonlinear and require numerical solutions (see Cherukuri and Johnson, 1996).

First, we note that the Equations (29) and (30) together with Equations (28) and (32) imply that

$$S_{xx} = -S_{yy} \quad \text{and} \quad \Omega = [S_{xx}^2 + \beta^2 S_{xy}^2]^{1/2} \tag{38}$$

Further through simple algebraic manipulations it can be shown that τ_s in Equation (35) is given by

$$\tau_s = S_{xy}(x, H^\pm(x, t), t) - 2\frac{\delta}{\beta} h'(x) S_{xx}(x, H^\pm(x, t), t) + \cdots \tag{39}$$

and that the sheet speed at the interface entering into Equation (35) is given by

$$u_s = u(x, H^\pm(x, t), t) + O(\delta^2). \tag{40}$$

It is also convenient to integrate the conservation of mass equation (28) across the roll-bite gap. Using Leibniz's formula, along with the kinematic boundary conditions (37), we obtain

$$2\varepsilon\frac{db}{dt} + \frac{\partial q}{\partial x} = 0, \tag{41}$$

where the volume flux q is

$$q(x, t) = \int_{H^-}^{H^+} u(x, y, t)dy. \tag{42}$$

Next, we note that since $\delta\beta \ll 1$, (31) implies that at the leading-order, $u(x, y, t)$ is independent of y. Therefore the volume flux given by (42) becomes

$$q(x, t) = u^{(0)}(x, t)(H^+(x, t) - H^-(x, t)) = 2u^{(0)}(x, t)H(x, t).$$

Substituting this in (41) and noting that the prescribed dimensionless flux is $q(a(t), t) = 2$, we get

$$u^{(0)}(x, t) = \frac{1}{H(x, t)} - \varepsilon\dot{b}(t)\frac{x - a(t)}{H(x, t)}. \tag{43}$$

The equilibrium equations (26) and (27) suggest that the pressure and deviatoric stress components can be expanded in an asymptotic series of the form

$$p(x, y, t) = p^{(0)}(x, y, t) + \delta^2 p^{(1)}(x, y, t) + \cdots,$$

$$S_{xx}(x, y, t) = -S_{yy}(x, y, t) = S_{xx}^{(0)}(x, y, t) + \delta^2 S_{xx}^{(1)}(x, y, t) + \cdots,$$

and

$$S_{xy}(x, y, t) = S_{xy}^{(0)}(x, y, t) + \delta^2 S_{xy}^{(1)}(x, y, t) + \cdots.$$

We will be concerned here with only the leading-order solution as it is accurate to $O(\delta^2)$. Substituting the above expansions in Equation (27), we obtain

$$-\frac{\partial}{\partial y}\left(p^{(0)} - \frac{\delta}{\beta}S_{yy}^{(0)}\right) = 0,$$

which enables us to define a function $N(x, t)$ such that

$$N(x, t) = p^{(0)} - \frac{\delta}{\beta}S_{yy}^{(0)} = p^{(0)} + \frac{\delta}{\beta}S_{xx}^{(0)}, \tag{44}$$

where we have used $S_{yy} = -S_{xx}$ from Equation (38). Note that $N(x, t)$ is essentially the leading-order terms in the vertical stress σ_{yy}. Substituting the above in Equation (26), we find that

$$\frac{\partial N}{\partial x}(x, t) = 2\frac{\delta}{\beta}\frac{\partial S_{xx}^{(0)}}{\partial x}(x, y, t) + \frac{\partial S_{xy}^{(0)}}{\partial y}(x, y, t). \tag{45}$$

From Equations (35), (39), and (40) we note that $S_{xy}^{(0)}(x, t), t)$ is equal to

$$S_{xy}^{(0)}(x, H^{\pm}(x, t), t) = \pm\frac{U^{(0)} - u^{(0)}(x, t)}{\Delta U} \pm 2\frac{\delta}{\beta}h'(x)S_{xx}^{(0)}(x, H^{\pm}(x, t), t). \tag{46}$$

Next we note that at leading order (29) implies that $S_{xx}^{(0)}$ is independent of y, namely

$$S_{xx}^{(0)}(x, t) = \left|\frac{\partial u^{(0)}}{\partial x}\right|^{-1+1/n}\frac{\partial u^{(0)}}{\partial x}. \tag{47}$$

Integrating Equation (45) from $y = H^-(x, t)$ to $H^+(x, t)$ and making use of Equations (46) and (47), we find that

$$\frac{\partial N}{\partial x} = \frac{1}{H(x, t)}\frac{U^{(0)} - u^{(0)}(x, t)}{\Delta U} + \frac{\delta}{\beta}\frac{2h'(x)}{H(x, t)}S_{xx}^{(0)}(x, t) + 2\frac{\delta}{\beta}\frac{\partial S_{xx}^{(0)}}{\partial x}(x, t), \tag{48}$$

which yields

$$N(x, t) = \frac{1}{\Delta U} \int \frac{U^{(0)} - u^{(0)}(x, t)}{H(x, t)} dx$$

$$+ 2\frac{\delta}{\beta} \left\{ [S_{xx}^{(0)}(x, t)]_{a(t)}^{x} + \int_{a(t)}^{x} \frac{h'(x)}{H(x, t)} S_{xx}^{(0)}(x, t)dx \right\} + N(a(t), t), \quad (49)$$

where the integration constant $N(a(t), t)$ is determined by the entry tension (Equation (34)). Noting that $\sigma_{xx} = -p^{(0)} + \frac{\delta}{\beta}S_{xx}^{(0)} = -N(x) + 2\frac{\delta}{\beta}S_{xx}^{(0)}$, the first of Equation (34) gives $N(a(t)) = -\sigma^{\text{entry}}(t) + 2\frac{\delta}{\beta}S_{xx}^{(0)}(a,(t), t)$ and therefore

$$N(x, t) = \frac{1}{\Delta U} \int_{a(t)}^{x} \frac{U^{(0)} - u^{(0)}(x, t)}{H(x, t)} dx$$

$$+ 2\frac{\delta}{\beta} \left\{ S_{xx}^{(0)}(x, t) + \int_{a(t)}^{x} \frac{h'(x)}{H(x, t)} S_{xx}^{(0)}(x, t)dx \right\}. \quad (50)$$

It is worth pointing out here that the term involving $S_{xx}^{(0)}$ in the last integral above was assumed to be small and hence neglected by Johnson (1994). However, here we retain this term for the calculation of spring and damping coefficients. The total vertical force per unit length on the work-rolls is then given by

$$\hat{F}^{(s)}(t) = \frac{1}{\delta} \hat{p}_0 \hat{l} \int_{a(t)}^{l} N dx \simeq \frac{1}{\delta} \hat{p}_0 \hat{l} \int_{0}^{l} N dx - N(0, t)a(t). \quad (51)$$

In the following we consider only a linear reduction model where the arc of contact between the rolls and the sheet is approximated by

$$h(x) = 1 - \Delta h x. \quad (52)$$

The case of a circular reduction was considered by Johnson (1944), and the interested reader is referred to this paper for more details. The details of the arc of contact are not important (see Cherukuri and Johnson, 1996), since we are interested in only the net force that the sheet exerts on the work-rolls. However, the linear reduction results in relatively simple expressions without sacrificing accuracy. For the linear reduction case it is easy to see from geometry that

$$a(t) = \frac{\varepsilon b(t)}{\Delta h}. \quad (53)$$

Further, for small vibrations, it is clear that the dynamic component of the change in the length of arc of contact between the work-rolls and the sheet is approximately equal to $a(t)$.

Since ε is small, we note that

$$\int_\varepsilon^x f(x)dx \simeq \int_0^x f(x)dx - \varepsilon f(0) \quad \text{and} \quad [H(x,t)]^{-1} \simeq [h(x)]^{-1}\left\{1 - \varepsilon\frac{b(t)}{h(x)} + O(\varepsilon^2)\right\}.$$

Further $S_{xx}^{(0)}(x,t)$ (Equation (47)) can be approximated by

$$S_{xx}^{(0)} \simeq \left[\frac{-h'(x)}{h^2(x)}\right]^{1/n}\left[1 - \frac{\varepsilon}{n}\left\{\frac{2}{h(x)}b(t) + \left(x - \frac{h(x)}{h'(x)}\right)\dot b(t)\right\}\right] + O(\varepsilon^2). \quad (54)$$

Substituting these approximations into Equation (50) and retaining terms only up to $O(\varepsilon)$, we find that

$$N(x,t) \simeq [N_s(x) - \sigma_s^{\text{entry}}] + \varepsilon b(t)[N_1(x) - C_1 + C_2 + N_3(x)]$$
$$+ \varepsilon\dot b(t)[N_2(x) + N_4(x)] - \varepsilon\sigma_D^{\text{entry}}(t), \quad (55)$$

with

$$C_1 = \frac{U-1}{\Delta U\Delta h} \quad \text{and} \quad C_2 = \left\{\frac{-h'(0)}{\Delta h}\right\}^{1+(1/n)} \quad (56)$$

$$N_s(x) = \int_0^x \frac{Uh(x)-1}{h^2(x)}dx + 2\frac{\delta}{\beta}\left[\frac{-h'(x)}{h^2(x)}\right]^{1/n}, \quad (57)$$

$$N_1(x) = \frac{1}{2\Delta U}\int_0^x \frac{2-Uh(x)}{h^3(x)}dx - \frac{4}{nh(x)}\frac{\delta}{\beta}\left[\frac{-h'(x)}{h^2(x)}\right]^{1/n}, \quad (58)$$

$$N_2(x) = \frac{1}{\Delta U}\int_0^x \frac{x}{h^2(x)}dx - \frac{2}{n}\frac{\delta}{\beta}\left(x - \frac{h(x)}{h'(x)}\right)\left[\frac{-h'(x)}{h^2(x)}\right]^{1/n}, \quad (59)$$

$$N_3(x) = 2\frac{\delta}{\beta}\left(1 + \frac{2}{n}\right)\int_0^x \left[\frac{-h'(x)}{h^2(x)}\right]^{1/n}\frac{dx}{h(x)}, \quad (60)$$

$$N_4(x) = \frac{2}{n}\frac{\delta}{\beta}\int_0^x \left(x - \frac{h(x)}{h'(x)}\right)\left[\frac{-h'(x)}{h^2(x)}\right]^{1/n}dx. \quad (61)$$

The terms in the first set of square brackets in (55) are the well-known steady state terms, and the rest are new dynamic contributions to the roll-pressure. Further, for convenience, the entrance strip tension has been expressed as the sum of a steady state part and a dynamic part. The total force per unit length that the sheet exerts on the work rolls is obtained by integrating (55) through the roll-bite

$$\hat F^{(s)}(t) = \delta^{-1}\hat\tau_0\hat l\int_a^1 N(x,t)dx = \delta^{-1}\hat\tau_0\hat l\left\{\int_0^1 N(x,t)dx - \frac{\varepsilon b(t)}{\Delta h}N(0,t)\right\} + O(\varepsilon^2).$$
$$(62)$$

Substituting (55) in Equation (62) and neglecting terms of $O(\varepsilon^2)$, the total sheet force per unit length can be expressed as

$$\hat{F}^{(s)}(t) = \delta^{-1}\hat{\tau}_0\hat{l}\{F_S^{(s)} + F_D^{(s)}(t)\}, \tag{63}$$

where

$$F_S^{(s)} = \int_0^1 N_s(x)dx - \sigma_s^{\text{entry}}, \tag{64}$$

and

$$F_D^{(s)}(t) = -\varepsilon K^{(s)}b(t) - \varepsilon S^{(s)}\dot{b}(t) - \varepsilon\sigma_D^{\text{entry}}(t), \tag{65}$$

where

$$K^{(s)} = -\int_0^1 \{N_1(x) + N_3(x)\}dx + C_1 - C_2 + \frac{1}{\Delta h}\left[2\frac{\delta}{\beta}(-h'(0))^{1/n} - \sigma_s^{\text{entry}}\right], \tag{66}$$

$$S^{(s)} = -\int_0^1 \{N_2(x) + N_4(x)\}\,dx. \tag{67}$$

Thus we note that $K^{(s)}$ denotes the effective spring coefficient and $S^{(s)}$ the effective damping coefficient for the interaction between the sheet and the work-rolls. Further it is also appropriate to mention here that the terms involving $a(t)$ (and therefore time dependent of the length of the arc of contact; see Equation (53)) contribute only to the spring coefficient $K^{(s)}$. This is in contrast to the conclusions by Tamiya et al. (1977) and Tlusty et al. (1982) that the periodic variation of the contact length at the exit of the strip contributes to damping.

Next we rescale the sheet force according to the nondimensionalization introduced in Section 4.2. Recalling that $\hat{Y}^i$ (and hence $\hat{b}(t)$) is nondimensionalized by $2\hat{F}_S(1 - v^2)\hat{E}^{-1}$ and the force by $\hat{F}_S$, we can write the dimensional total sheet force as

$$\hat{F}^{(s)}(t) = \delta^{-1}\hat{\tau}_0\hat{l}F_S\left[1 + \frac{\hat{F}_D^{(s)}(t)}{\hat{F}_S}\right] = \hat{F}_S[1 + F_D^{(s)}(t)],$$

where $F_D^s(t)$ is now the dynamic sheet-force nondimensionalized by $\hat{F}_S$ and given by

$$F_D^{(s)}(t) = -\varepsilon[K^{(s)}\eta B(t) + S^{(s)}\eta\alpha\dot{B}(t) + \phi\sigma_D^{\text{entry}}(t)], \tag{68}$$

where the constants η, α, and ϕ are defined as

$$\eta = \frac{2(1-v^2)\hat{\tau}_0\hat{l}}{\pi\hat{E}\hat{h}_0\delta}, \quad \alpha = \frac{\hat{l}}{\hat{u}_0}\sqrt{\frac{\pi\hat{E}}{2\hat{m}(1-v^2)}}, \quad \text{and} \quad \phi = \frac{\hat{\tau}_0\hat{l}}{\delta\hat{F}_S}. \tag{69}$$

It is worth pointing out that the same symbol $F_D^{(s)}(t)$ is used to denote the dynamic part of the sheet force in Equations (65) and (68) where the nondimensionalization is different in each case. This is not expected to cause confusion, since the quantity of interest to us in analyzing the vibration is that given by Equation (68). However, we note that the spring and damping coefficients are the same as defined in Equations (66) and (67) for both versions of the nondimensionalization.

The parameter η is the ratio of two length scales: the magnitude of elastic displacement due to the characteristic force in the roll-bite and the sheet thickness. The parameter α is the ratio of the residence time of the material in the bite to the time scale associated with the elastic restoring force.

Until now we assumed that the roll-speed U is known. However, in practice, it is popular to replace the roll-speed U by the slip f defined by

$$f = \frac{u_1 - U}{U} \quad \text{or} \quad U = \frac{u_1}{1+f} = \frac{1}{(1-\Delta h)(1+f)}, \tag{70}$$

where u_1 is the exit-speed of the sheet under steady state rolling conditions. In Figures 4.8 and 4.9, the spring and damping coefficients $K^{(s)}$ and $S^{(s)}$ are plotted as functions of reduction and δ/β for $n=4$ and $n=100$. The slip is taken to be 0.15 and $\Delta U = 1$. Note that $\Delta U = 1$ implies that the characteristic velocity is taken as $\Delta\hat{U} = \hat{u}_0$, the entry sheet speed (or the processing rate). First, we note that for a fixed δ, increasing δ/β corresponds to decreasing friction, decreasing processing rates $\hat{u}_0$, or decreasing roll-bite length. The effect of friction on spring coefficient $K^{(s)}$ is evident from Figure 4.8. As δ/β decreases and therefore as friction increases, $K^{(s)}$ decreases. In fact, for small enough reductions and large enough friction, $K^{(s)}$ can become negative. Further, comparison of Figure 4.8a with 4.8b indicates that for moderate reductions of the order 40% or less, larger n yields a larger spring coefficient. However, $K^{(s)}$ is less sensitive to n as the reductions get larger. Further it is evident from Equation (66) that the entry tension σ_s^{entry} can drive $K^{(s)}$ down and, when sufficiently large, can even make $K^{(s)}$ negative. However, as discussed earlier, the total spring coefficient for the vibrations of the rolls consists of a contribution from the elastic deformation at the interface of the backup and work-rolls, in addition to $K^{(s)}$. Thus whether or not the total spring coefficient is negative depends on the size of the contribution coming from the backup rolls.

The effect of friction on damping coefficient $S^{(s)}$ is evident from Figure 4.9. As friction increases, $S^{(s)}$ decreases. For large enough friction and reduction, negative damping can occur when the rate-sensitivity index n is small. How-

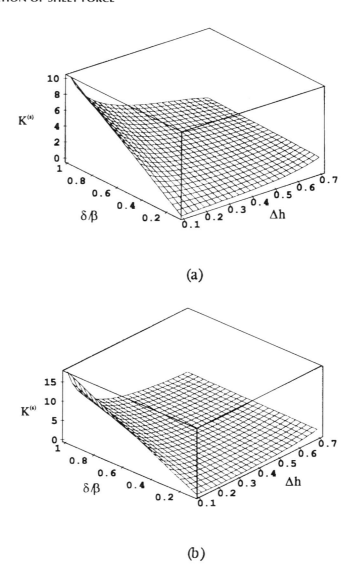

(a)

(b)

FIGURE 4.8. Effective spring coefficient $K^{(s)}$ due to the sheet deformation in the roll-bite: (a) $n = 4$; (b) $n = 100$.

ever, as n gets bigger (i.e., for weakly rate-sensitive materials), the damping coefficient can be negative for the entire range of reduction and friction levels, as shown in Figure 4.9b. Since the entire damping in the roll-mill system comes from sheet deformation, it is clear that negative damping coefficients are to be avoided to prevent chatter.

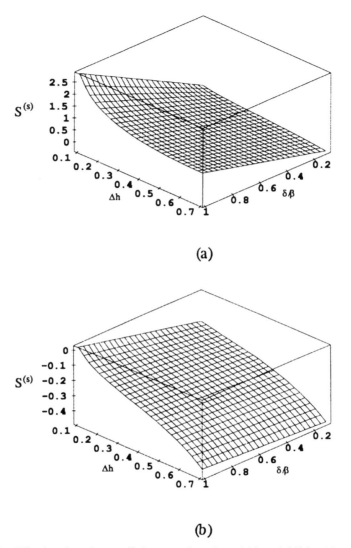

(a)

(b)

FIGURE 4.9. Effective damping coefficient as a function of Δh and δ/β for (a) $n = 4$ and (b) $n = 100$.

The sources of negative damping and spring coefficients can be traced to the terms that have a negative contribution to $F_D^{(s)}$ in Equation (68). Following the physical explanation given in (1994), we note that an increase in the gap between the work-rolls leads to a decrease in the sheet speed through the gap and consequently to an increase in the velocity difference between the sheet and the rolls, which increases the shear stress at the roll-sheet interface. Since shear stresses drive the pressure in the bite, this will increase the pressure and therefore lead to a greater force on the rolls. The additional force drives the rolls to separate further, and consequently damping or spring coefficients become negative. However, this behavior is critically dependent on the nature of the friction model used and on the magnitude of friction. Consequently friction plays an important role in chatter phenomena.

4.8 CONCLUSION

A formulation of 2-DOF and 4-DOF systems has been presented to study chatter phenomenon in rolling-mills. Both the 2-DOF and 4-DOF systems consist of a squeezing mode and a motion of the mass-center. The nonlinear force-displacement relation between the work- and backup rolls is linearized, and closed-form expressions for the natural frequencies are obtained for both systems. In addition to the elastic deformation between the work- and backup rolls, the plastic deformation of the sheet also contributes to the net spring coefficient of the squeezing mode vibrations. The mass-center motion is unaffected by the dynamic motion of the sheet; that is the motion is damping free and the spring coefficient depends only on the elastic deformation at the interfaces of work- and backup rolls. As friction increases (i.e., as δ/β decreases), the contribution of the sheet deformation to the net spring coefficient of the squeezing motion decreases. Consequently, when friction is sufficiently large, the natural frequencies for the 2-DOF system indicate that the squeezing mode frequency differs slightly from the mass-center frequency. When the two frequencies are close and damping is negligible, beating phenomenon is possible. Further, for the numerical values tested here, these frequencies correspond to fifth-octave chatter vibrations. For the 2-DOF system the nonlinear force-displacement relation at the interface between the work- and backup rolls, first identified by Roberts (1978), is found to trigger higher harmonics in the vibration, but it causes only a small frequency shift.

For the 4-DOF system the eigen-modes indicate that the motion consists of two squeezing modes and two mass-center motions. The squeezing modes lead to gauge variations, whereas the mass-center motions lead to sheet corrugations. The two squeezing modes seem to correspond roughly to the third- and fifth-octave chatter vibrations. The mass-center motions correspond to third-octave chatter modes.

Analytical expressions for spring and damping coefficients due to the plastic deformation of the sheet are obtained by a regular perturbation approach where the thinness parameter δ is assumed to be small. In particular, the case of a linear reduction in the roll-bite gap is considered in detail. The precise nature of the reduction is not of importance, since we are interested in effective spring and damping coefficients that are global quantities. However, the interested reader is referred to Johnson (1994) for details on the circular reduction case.

The spring and damping coefficients due to the sheet deformation depend on various factors such as the reduction, entrance gauge, roll-bite length, roll-sheet friction, processing rate, and material behavior. In addition, at the leading-order, only the spring coefficient depends on the entrance tension and the roll-speed. As the shear or friction increases (i.e., as δ/β decreases), the spring coefficient due to sheet deformation, $K^{(s)}$, decreases. Because the characteristic shear stress is $\hat{\tau}_0 = \hat{\kappa}\hat{u}_0$, increasing the friction corresponds to higher processing rates or a higher friction coefficient $\hat{\kappa}$. Also, as the slip f increases, $K^{(s)}$ decreases. This is because larger slips imply larger relative velocities between the sheet and the rolls and therefore greater friction. In particular, if slip is sufficiently large or friction is sufficiently high, $K^{(s)}$ can become negative. For small reductions and $\delta/\beta \cong 1$, the spring coefficient is very sensitive to the rate-sensitivity parameter n. However, as n increases or as δ/β decreases, the spring coefficient becomes less sensitive to n. The entrance tension also decreases $K^{(s)}$ and can make it negative. However, the net spring coefficient that includes the contribution from the elastic deformation between the backup and work-rolls may still be positive. When friction is small, $K^{(s)}$ decreases with reduction, and the change can be substantial. However, for large enough friction, $K^{(s)}$ starts off from a negative value for small reductions, and as reduction increases, the spring coefficient also increases and becomes positive.

The damping coefficient $S^{(s)}$ also decreases with increasing friction or processing rate. Whether damping is positive or negative depends strongly on the rate-dependence parameter n. As the rate-dependence becomes weak (i.e., as n gets larger), damping can become negative, and when n is large enough, damping can be negative for all values of δ/β and reduction. Thus, in weakly rate-dependent materials common in cold rolling, negative damping is highly likely, and consequently self-sustained vibrations are possible. Negative damping is also possible for large reductions and large friction. Further damping is a monotonically decreasing function of reduction, with the lowest damping occurring at the highest reduction. The observation that damping decreases with increasing friction is consistent with the conclusion made by Yarita et al. (1978) that degeneration of lubricant (and hence increased friction) leads to chatter.

The variation in the contact length has been shown to contribute only to the spring coefficient and not to the damping coefficient. This is in contrast to the conclusions reached by Tamiya et al. (1977) and Tlusty et al. (1982) that the periodic variations in the length of the arc of contact contribute to positive damping and thus to self-sustained vibrations. Further we have shown that the effect of the spring coefficient due to sheet deformation $K^{(s)}$ may be small and insignificant for some values of reduction and friction level, but the damping coefficient $S^{(s)}$ (due entirely to the plastic deformation of the sheet) is not necessarily insignificant. In fact, when rate-dependence is important (i.e., for hot rolling problems), $S^{(s)}$ is positive and significant enough to damp out the vibrations of the rolls very quickly. On the other hand, when rate-dependence is not very important (i.e., for cold-rolling problems), $S^{(s)}$ is negative and leads to chatter. Again, this is in contrast to several previous studies that assume the effect of the sheet on the overall vibrational characteristics of the rolling-mill system to be negligible.

In conclusion, this study shows the importance of friction and rate-sensitivity in roll-chatter dynamics. Chatter is more likely to occur in weakly rate-sensitive materials or when reduction and friction are high, since in both the cases damping can become negative. Further sufficiently large entry tension can also make the spring coefficient due to the sheet-deformation become negative, and hence it is possible for the total spring coefficient to be negative. Thus interstand tension can lead to negative spring coefficients and therefore to chatter. Again, this conclusion is different from those of several previous studies where interstand tension is claimed to lead to negative damping.

REFERENCES

Chefneux, L., and Gouzou, J. 1984. Study and industrial control of chatter in cold-rolling. *Iron Steel Eng.* **61**:17–26.

Cherukuri, H. P., Johnson, R. E., and Smelser, R. E. 1996. A rate-dependent theory for hot-rolling. *Int. J. Mech. Sci.*, **39**:705–727.

Guo, R., Urso, A. C., and Schunk, J. H. 1993. Analysis of chatter vibration phenomena of rolling mills using finite element methods. *Iron Steel Eng.* **70**:29–39.

Johnson, R. E. 1994. The effect of friction and inelastic deformation on chatter in sheet rolling. *Proc. R. Soc. London. A* **445**:479–499.

Johnson, R. E., and Qi, Q. 1994. Chatter dynamics in sheet rolling. *Int. J. Mech. Sci.* **36**:617–630.

Misonoh, K. 1980. Analysis of chattering in cold rolling of steel strip. *J. Jap. Soc. Tech. Plast.* **21**:1006–1010.

Nessler, G. L. 1989. Cause and solution of fifth-octave backup roll chatter on 4-h cold mills and temper mills. *Iron Steel Eng.* **66**:33–37.

Pawelski, O., Rasp, W., and Friedewald, K. 1986. Application of the theory of rolling to rolling in the case of mill vibrations. *Steel Res.* **57**:373–376.

Roark, R. J., and Young, W. C. 1975. *Formulas for Stress and Strain.* McGraw-Hill, New York.

Roberts, W. L. 1978. Four-h mill-stand chatter of the fifth-octave mode. *Iron Steel Eng.* **55**:41–47.

Roberts, W. L. 1988. *Flat Processing of Steel.* Marcel Dekker, New York.

Tamiya, T., Furui, K., and Iida, H. 1977. Analysis of chattering phenomena in cold-rolling. *Iron Steel Eng. Yearbook*, pp 1191–1207.

Tlusty, J., Chandra, G., Critchley, S., and Paton, D. 1982. Chatter in cold-rolling. *Ann. CIRP* **31**:195–199.

Yarita, I., Furukawa, K., Seino, Y., Takimoto, T., Nakazato, O., and Nakagawa, K. 1978. An analysis of chattering in cold rolling for ultra thin guage steel strip. *Trans. ISIJ* **18**:1–10.

5

MODELING, ANALYSIS, AND CHARACTERIZATION OF MACHINING DYNAMICS

I. MINIS and B. S. BERGER

5.1 INTRODUCTION

The dynamical characteristics of machining systems lead to the appearance of large amplitude motions between cutting tool and workpiece in the case of large material removal rates; this phenomenon is called *machining chatter*. Beyond limiting productivity, chatter has detrimental effects on workpiece quality as well as cutting tool and machine tool life. As a result over the last three decades substantial research on machining dynamics has targeted chatter prediction and control.

The dynamics of machining are governed by the dynamics of the machine tool, the dynamics of the cutting process, and the external and internal excitations, such as those generated by chip breaking, the built-up edge clearing away from the tool, or the nonhomogeneous hardness distribution of the material being cut. During cutting the structural and process dynamics are intrinsically coupled, as shown in Figure 5.1. The cutting force F excites the machine tool structure, which responds with a tool-workpiece relative displacement r. This in turn excites the process dynamics.

Dynamics and Chaos in Manufacturing Processes, Edited by Francis C. Moon.
ISBN 0-471-15293-5 © 1998 John Wiley & Sons, Inc.

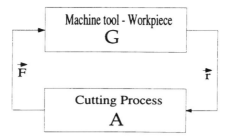

FIGURE 5.1. Closed-loop model of machining dynamics.

Although machine tool structures are complex distributed parameter systems, for the analysis of cutting dynamics it is sufficient to consider the relationship between a pair of opposite forces applied on the cutting tool and the workpiece, respectively, and the relative displacement that these forces induce. This relationship has traditionally been considered linear and represented in the frequency domain by a single transfer function or a third-order receptance matrix (e.g., see Koenigsberger and Tlusty, 1971; Weck, 1985). Well-established methods of structural dynamics have been employed to identify the structural transfer functions experimentally (Tobias, 1965; Taylor, 1977; Kim et al., 1984). It has been postulated and experimentally verified, however, that in some cases the structure-workpiece system exhibits intrinsic nonlinearities, which are usually modeled by nonlinear stiffness terms (Hanna and Tobias, 1969; Klamecki, 1989).

The cutting process dynamics depend on the values of the machining parameters, namely feed rate, spindle speed, depth of cut, tool geometry and consistency, and workpiece geometry and consistency. Most research work on the analysis of the cutting process has addressed orthogonal cutting in which the cutting edge is perpendicular to the direction of the feed. Although this geometry simplifies the analysis, the phenomena intrinsic to the cutting process, namely plastic deformation under high strain, strain hardening, strain rate sensitivity, friction between the chip and the tool rake face and chip curling, are extremely complex and not well understood to this day. The cutting process dynamics have been analyzed extending models of steady cutting to relate the cutting forces to the shear angle (Merchant, 1945). The dynamic variation of the latter has been examined closely and used to postulate nonlinear cutting force equations (Wu and Liu, 1985a, 1985b; Wu, 1988, 1989; Lin and Weng, 1990). The lack of adequate measures of nonlinear dynamical behavior and of appropriate signal-processing techniques has prevented in-depth experimental validation of these models. The experimental studies have focused instead on empirical linear cutting force models. Methods of time series analysis have been employed to estimate the parameters of linear transfer functions that best approximate input-output cutting measurements (Tlusty, 1978; Peters et al., 1971; Ahn et al., 1985). Nonlinear cutting equations have been combined with linear structural models to show that under certain conditions the coupled machining system may exhibit chaotic oscillations (Grabec, 1986, 1988).

The results of studies of structural and cutting process dynamics have been used for chatter prediction. The majority of prediction methods represent the interaction between the cutting process and the machine tool-workpiece system by a linear, closed-loop system with time delays. Well-established techniques of control theory are used to determine the system's stability limit, which corresponds to the critical depth of cut for chatter-free machining (Merritt, 1965; Weck, 1985). Nonlinear models of the metal-cutting system have also been used for chatter prediction. The resulting equations have been examined either numerically (Jemielniak and Widota, 1989; Tlusty and Ismail, 1983) or by extending the existing methods of linear analysis to treat simple nonlinearities (Shi and Tobias, 1984).

The methods employed to detect the imminence of chatter usually rely on quantitative changes of a certain output variable. For example, certain characteristics of the sound emitted by the cutting process have been related to both the state of the machining system and the corresponding cutting conditions. Delio et al. (1992) were successful in using audio signals recorded by microphones placed near the cutting zone to detect chatter in milling. An alternative technique is based on the measurement of the deflection of the work piece during machining (Rahman and Ito, 1986; Rahman, 1988). Experiments have shown that in certain cases the relationship between this quantity and the depth of cut exhibits characteristic patterns immediately prior to the onset of chatter. This criterion can be employed only if the structural dynamics are dominated by the modes of the main spindle drive and/or the workpiece itself (Rahman, 1988). An interesting approach for chatter detection uses stochastic identification techniques to fit linear ARMA models to on-line measurements of the cutting force and the workpiece acceleration (Eman and Wu, 1980). Critical system parameters are extracted from the resulting models and used as warning indicators for the imminence of chatter.

This chapter discusses prediction of chatter and characterization of machining dynamics from on-line measurements. For chatter prediction we present a linear model of machining dynamics. Section 5.1 discusses experimental methods for the identification of the receptance matrix of the machine tool structure and the development of the corresponding modal differential equations. Section 5.2 presents a set of semiempirical linear force equations that describe the cutting process dynamics, and discusses experimental ways to determine the equations' parameters. In Section 5.3 the structural model and the force relationships are used to develop a system of differential-difference equations that represents the dynamics of machining. The stability of the system's equilibrium point is related to machine-tool chatter. Standard stability methods are used to predict chatter in turning. In milling, the coefficients of the dynamical equations become periodic. For this case special stability analysis techniques are presented and applied to a small milling example.

Despite the intrinsic nonlinearity of machining dynamics, linear models and techniques are appropriate for chatter prediction. This is not true, however, for applications such as early chatter detection and control, which require a fundamental understanding of the phenomena involved. For this purpose, Section 5.4 presents the use of signal-processing techniques, motivated by nonlinear dynamics, to determine fundamental dynamical measures that can be potentially used to characterize the machining state.

5.2 STRUCTURAL DYNAMICS OF MACHINE TOOLS

This section presents two alternative linear representations of the machine tool structural dynamics. The first is a transfer-function-based model, which is an appropriate model for experimental identification of the structural dynamics and for chatter prediction via linear stability methods. The second is a time domain model that represents the structure's modal response.

5.2.1 The Transfer Function Model

The following relationship represents the dynamics of the machine tool structure at the cutting tool.

$$\mathbf{r}(j\omega) = G(j\omega)\mathbf{F}(j\omega),$$

$$\begin{bmatrix} x^{(m)}(j\omega) \\ y^{(m)}(j\omega) \\ z^{(m)}(j\omega) \end{bmatrix} = \begin{bmatrix} G_{xx}(j\omega) & G_{xy}(j\omega) & G_{xz}(j\omega) \\ G_{yx}(j\omega) & G_{yy}(j\omega) & G_{yz}(j\omega) \\ G_{zx}(j\omega) & G_{zy}(j\omega) & G_{zz}(j\omega) \end{bmatrix} \begin{bmatrix} F_x^{(m)}(j\omega) \\ F_y^{(m)}(j\omega) \\ F_z^{(m)}(j\omega) \end{bmatrix}, \tag{1}$$

where $\mathbf{r}(j\omega)$ is the vector of the Fourier transforms of the tool displacements and $\mathbf{F}(j\omega)$ is the vector of the Fourier transforms of the cutting force components applied on the cutting tool. $G(j\omega)$ is the receptance matrix of the machine tool structure. All these quantities are expressed in terms of the coordinate system (m) which is fixed with respect to the machine tool structure (see Figure 5.2). The elements $G_{ab}(j\omega)$ of matrix G are given by

$$G_{ab}(j\omega) = \frac{a_a^{(m)}(j\omega)}{F_b^{(m)}(j\omega)} = G_{ba}(j\omega), \tag{2}$$

where $a_a^{(m)}(j\omega)$ is the Fourier transform of the tool's response along direction a due to an excitation $F_b^{(m)}(j\omega)$ along direction b ($a, b = x^{(m)}, y^{(m)}$, or $z^{(m)}$). In general, the relative tool-workpiece displacements are affected by the vibrational dynamics of the workpiece. In this discussion, however, the workpiece is assumed to be rigid for simplicity. This assumption is valid for stubby workpieces. Tembo (1994) discusses the case of flexible workpieces.

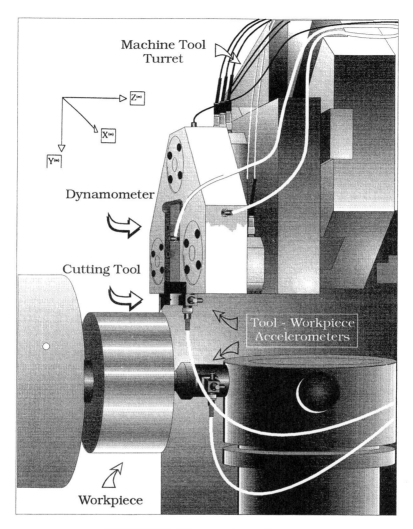

FIGURE 5.2. The experimental setup.

5.2.2 Experimental Determination of the Transfer Function

Two experimental techniques for determining the receptance matrix of Equation (1) are overviewed below. The first comprises a set of simple impulse tests performed when the machine tool is not cutting. The second technique is more sophisticated and identifies the receptance matrix under normal cutting conditions. This provides a better estimate of the structure's damping which is a critical factor in chatter prediction. We describe the application of both techniques on a Hardinge CNC lathe that is equipped with a special force dynamometer (employing three Kistler 9068 force transducers) and its associated electronics, accelerometers (Kistler 8628B50), and their electronics, and an eight-channel digital spectrum analyzer (Hewlett Packard 3566A). Figure 5.2 shows the experimental apparatus.

Impulse Tests

During these tests the turret of the lathe was lowered into the cutting position and was programmed to move at a constant feed rate $d_0 = 0.127$ mm/rev along the $-Z^{(m)}$-axis to simulate the turning process. Actual cutting was not performed, so as to uncouple the closed loop machining system of Figure 5.1. The turret of the lathe was excited with an impulse hammer (PCB 208A03) along the b-axis ($b = x^{(m)}, y^{(m)}, z^{(m)}$). The applied force $F_b^{(m)}$ was measured by the dynamometer, and the acceleration along the a-axis ($a = x^{(m)}, y^{(m)}, z^{(m)}$) was measured using accelerometers located at the cutting point. The accelerations and force measurements were recorded and analyzed by the HP3566A spectrum analyzer to obtain the frequency responses $G_{ab}(j\omega)$.

Figure 5.3 shows a typical receptance plot obtained from the impulse tests. The receptance G_{zz} exhibits peaks at $f_1 = 97$ Hz, $f_2 = 135$ Hz, and $f_3 = 260$ Hz, which correspond to the first three modes of the tool's structure. The dominant mode is at 97 Hz and the other two modes have much smaller

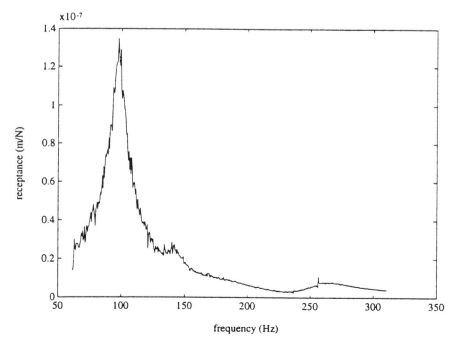

FIGURE 5.3. Magnitude of receptance G_{zz} obtained from impulse test.

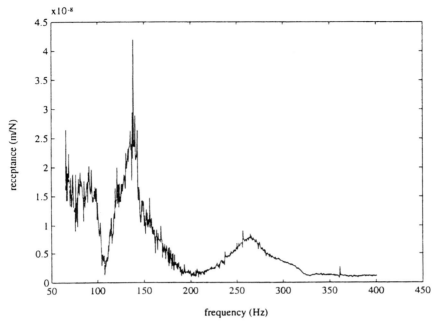

FIGURE 5.4. Magnitude of receptance G_{yz} obtained from impulse test.

contributions. The relative magnitudes of these peaks in other receptances G_{ab} vary. For example, Figure 5.4 shows that the second and third modes provide a major contribution to the receptance G_{yz}. This difference in relative magnitudes is due to the orientation of the modes with respect to the structure's coordinate system.

The receptances show that the three most prominent modes of the machine tool structure are well separated. Thus the parameters of each mode in each receptance G_{ab} may be estimated using a single-degree-of-freedom curve-fitting procedure (Ewins, 1984). In this case the influence of the other two modes is modeled by a complex constant that is added to the frequency response.

The portion of the receptance G_{ab} in the neighborhood of the resonant frequency f_i is designated as G_{abi}. Figure 5.5 shows the fit between the measurements and the single-degree-of-freedom model for G_{xx1} in the Nyquist plane. The parameters of the single-degree-of-freedom models for the three significant modes of the Hardinge lathe and the six elements of the receptance matrix are summarized in Table 5.1. The cells with no entries indicate that the corresponding modes did not contribute significantly to the receptances.

Cutting Tests

Many researchers have observed that for precise identification of the structural dynamics, the state of the machining system must simulate, as closely as

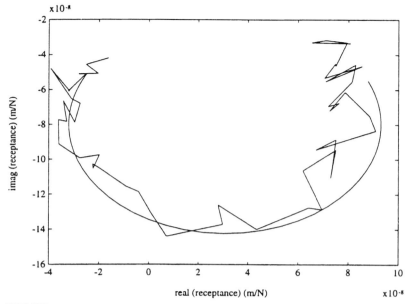

FIGURE 5.5. Nyquist plot of G_{xx1} and the fitted single-degree-of-freedom model.

TABLE 5.1
Parameters of Receptances G_{ab}

G_{ab}	Mode 1	Mode 2	Mode 3
G_{xx}	$m_{xx1} = 99.1\,\text{kg}$ $f_{xx1} = 96.4\,\text{Hz}$ $b_{xx1} = 0.102$	$m_{xx2} = 211.9\,\text{kg}$ $f_{xx2} = 132.3\,\text{Hz}$ $b_{xx2} = 0.06$	
G_{yx}	$m_{yx1} = 151.9\,\text{kg}$ $f_{yx1} = 97.5\,\text{Hz}$ $b_{yx1} = 0.077$	$m_{yx2} = 417.2\,\text{kg}$ $f_{yx2} = 136.5\,\text{Hz}$ $b_{yx2} = 0.098$	$m_{yx3} = 368.7\,\text{kg}$ $f_{yx3} = 262.0\,\text{Hz}$ $b_{yx3} = 0.064$
G_{zx}	$m_{zx1} = 151.9\,\text{kg}$ $f_{zx1} = 97.5\,\text{Hz}$ $b_{zx1} = 0.077$		
G_{yy}		$m_{yy2} = 208.8\,\text{kg}$ $f_{yy2} = 123.2\,\text{Hz}$ $b_{yy2} = 0.093$	$m_{yy3} = 98.5\,\text{kg}$ $f_{yy3} = 236.2\,\text{Hz}$ $b_{yy3} = 0.017$
G_{yz}		$m_{yz2} = 621.6\,\text{kg}$ $f_{yz2} = 135.7\,\text{Hz}$ $b_{yz2} = 0.041$	$m_{yz3} = 373.2\,\text{kg}$ $f_{yz3} = 259.8\,\text{Hz}$ $b_{yz3} = 0.054$
G_{zz}	$m_{zz1} = 154.8\,\text{kg}$ $f_{zz1} = 97.1\,\text{Hz}$ $b_{zz1} = 0.072$		

Source: Tembo (1994).

possible, the cutting state (Moriwaki and Iwata, 1976). However, the intrinsic coupling of the machining system during cutting (see Figure 5.1) presents severe problems for the identification of the structure from input-output measurements. Thus identification during cutting is only possible in special system configurations. Such a case can be realized by interrupted cutting. Consider, for example, the turning of the special work piece shown in Figure 5.6. In this case the motion of the tool between consecutive engagements with the teeth of the workpiece is a free vibration, and therefore the equations describing the cutting process would be valid for the time intervals when cutting is performed. During the free motion of the tool, the applied force components are zero. Consequently the excitation applied on the tool not only is now a function of the chip parameters, but it also depends on time explicitly. This time dependence is dictated by the geometry of the workpiece and the cutting speed. It has been shown by both Fisher (1965) and Gustavsson, et al. (1977) that if the feedback element of a closed-loop system is time varying (or nonlinear), then identification can be performed by input-output experiments disregarding the fact that the data were collected during closed loop operation.

The workpiece of Figure 5.6 was used in Minis et al. (1990a) to generate a broadband-cutting force signal. The pattern of channels in the axial direction of the workpiece created a pseudorandom profile, namely a discrete-interval binary sequence of teeth (1) and channels (0) that changes state only at specified integer multiples of an elementary length Δl. Figure 5.7 shows the power spectrum of $F_x^{(m)}$ generated by interrupted, orthogonal facing of the pseudorandom workpiece with 255 linear lengths at a rotational speed of 240 rpm. At this speed the resolution of the spectrum is 4 Hz, and the bandwidth of the first lobe is 1020 Hz. The lower and upper limits of the usable identification bandwidth are approximately 19 and 510 Hz, respectively.

Note that during orthogonal cutting, two cutting force components (e.g., $F_z^{(m)}$ and $F_x^{(m)}$) excite the structure simultaneously. Thus the tool displacement along each direction is the superposition of the structure's responses due to both forces. Furthermore the exciting forces are linearly dependent. Both these facts present additional difficulties in the identification of the receptance elements. Minis et al. (1990a) present a method that determines pairs of receptance elements by combining the measurements obtained from two specially chosen cutting configurations. Having determined all receptances, single-degree-of-freedom models may be evaluated following the procedure outlined above.

When the modal parameters from the impulse and cutting tests are compared, it is found that oftentimes the impulse method underestimates the damping of the structure. For example, for the receptance element G_{zz} of the Hardinge AB lathe, the impulse method yielded a damping ratio of 0.103, while the cutting method yielded a value of 0.127. This increase in damping during cutting may be attributed to many factors, such as the pressure increase of the lubricating oil of the machine tool slides or the difference in the magnitude of the exciting force in the two methods. It is noted that since the accurate

FIGURE 5.6. An aluminum pseudorandom work piece used for interrupted cutting. (From Minis et al., 1990a.)

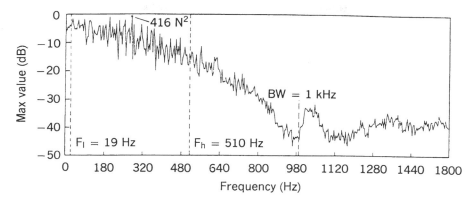

FIGURE 5.7. Power spectrum of the cutting force component $F_x^{(m)}$ generated by interrupted cutting. (From Minis et al., 1990a.)

estimation of the damping is critical in the analysis of machine tool chatter, the results of the impulse method should be used with caution.

5.2.3 Determination of the Structural Modes

Many applications use the time domain representation of the machine tool structural dynamics. For instance, a time-domain model is appropriate for the analysis of machining dynamics when nonlinear cutting force models are employed. In addition time-domain models are most useful in machine tool design. To construct such models, the identification of the most significant structural modes and their directions is necessary. Tembo (1994) presents the method outlined below to determine the structural modes from the experimentally obtained receptances $G_{ab}(j\omega)$.

Let the effect of each mode of the machine tool structure at the cutting point be represented by an oscillator (see Figure 5.8). The orientation of this oscillator with respect to the machine coordinate system is defined by the angles ϕ_i (between the $Z^{(m)}$-axis and the direction ε_i in the $X^{(m)}-Z^{(m)}$ plane) and ψ_i (between the direction ε_i and the axis of the oscillator q_i in the $\varepsilon_i - Y^{(m)}$ plane). The equation of motion of the oscillator of Figure 5.8 is given by

$$m_i\ddot{q}_i + b_i\dot{q}_i + k_iq_i = F_x^{(m)}\sin\phi_i\cos\psi_i + F_y^{(m)}\sin\psi_i + F_z^{(m)}\cos\phi_i\cos\psi_i, \quad (3)$$

where m_i, b_i, and k_i represent the mass, damping, and stiffness, respectively, of mode i at the cutting point and q_i represents the displacement of the oscillator. $F_x^{(m)}$, $F_y^{(m)}$, and $F_z^{(m)}$ are the cutting forces in the $X^{(m)}$, $Y^{(m)}$, and $Z^{(m)}$ directions, respectively. The corresponding displacements along these directions are given

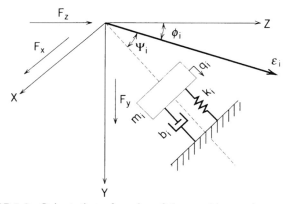

FIGURE 5.8. Orientation of modes of the machine tool structure.

by

$$x_i^{(m)} = q_i \cos \psi_i \sin \phi_i,$$
$$y_i^{(m)} = q_i \sin \psi_i,$$
$$z_i^{(m)} = q_i \cos \psi_i \cos \phi_i. \tag{4}$$

From the first of Equation (4), q_i can be expressed in terms of $x_i^{(m)}$. Using this relationship for q_i in Equation (3), taking Fourier transforms, and solving for $x_i^{(m)}$ yields

$$x_i^{(m)}(j\omega) = \frac{\sin^2 \phi_i \cos^2 \psi_i}{k_i - m_i\omega^2 + jb_i\omega} F_x^{(m)}(j\omega) + \frac{\sin \psi_i \sin \phi_i \cos \psi_i}{k_i - m_i\omega^2 + jb_i\omega} F_y^{(m)}(j\omega)$$
$$+ \frac{\cos \phi_i \sin \phi_i \cos^2 \psi_i}{k_i - m_i\omega^2 + jb_i\omega} F_z^{(m)}(j\omega) \tag{5}$$

Also

$$x^{(m)} = \sum_{i=1}^{v} x_i^{(m)}, \tag{6}$$

where v is the number of modes that contribute significantly to the dynamics of the structure.

Hence

$$x^{(m)}(j\omega) = \sum_{i=1}^{v} \frac{\sin^2 \phi_i \cos^2 \psi_i}{k_i - m_i\omega^2 + jb_i\omega} F_x^{(m)}(j\omega) + \sum_{i=1}^{v} \frac{\sin \psi_i \sin \phi_i \cos \psi_i}{k_i - m_i\omega^2 + jb_i\omega} F_y^{(m)}(j\omega)$$
$$+ \sum_{i=1}^{v} \frac{\cos \phi_i \sin \phi_i \cos^2 \psi_i}{k_i - m_i\omega^2 + jb_i\omega} F_z^{(m)}(j\omega). \tag{7}$$

Comparing Equatons (1) and (7), it is clear that the receptances G_{xb} ($b = x, y, z$) can be expressed as a function of the parameters m_i, b_i, k_i, ϕ_i, and ψ_i:

$$G_{xx} = \sum_{i=1}^{v} \frac{\sin^2 \phi_i \cos^2 \psi_i}{k_i - m_i \omega^2 + jb_i \omega},$$

$$G_{xu} = \sum_{i=1}^{v} \frac{\sin \psi_i \sin \phi_i \cos \psi_i}{k_i - m_i \omega^2 + jb_i \omega}, \qquad (8)$$

$$G_{xz} = \sum_{i=1}^{v} \frac{\cos \phi_i \sin \phi_i \cos^2 \psi_i}{k_i - m_i \omega^2 + jb_i \omega}.$$

Similar relationships can be obtained for G_{yb} and G_{zb} ($b = x^{(m)}, y^{(m)}, z^{(m)}$) using the remaining two Equations (4) and repeating the procedure.

Using the first of Equations (8) the following quantities can be defined,

$$m_{xxi} \equiv \frac{m_i}{\sin^2 \phi_i \cos^2 \psi_i},$$

$$k_{xxi} = m_{xxi} \omega_{xxi}^2 \equiv \frac{k_i}{\sin^2 \phi_i \cos^2 \psi_i}, \qquad (9)$$

$$b_{xxi} \equiv \frac{b_i}{\sin^2 \phi_i \cos^2 \psi_i},$$

where m_{xxi} is the modal mass, ω_{xxi} is the frequency of resonance, and b_{xxi} is the damping of the ith mode appearing in the receptance G_{xx}. Similar relationships can be used for G_{ab} ($a, b = x^{(m)}, y^{(m)}, z^{(m)}$).

Using Equations (9), their counterparts and the values of the parameters m_{abi}, b_{abi}, and k_{abi} ($a, b = x^{(m)}, y^{(m)}, z^{(m)}; i = 1, \ldots, v$) of the receptances (see Table 5.1), the values of the modal parameters m_i, b_i, k_i and the modal directions ϕ_i and ψ_i can be determined for each significant mode i. For the Hardinge lathe, Table 5.2 shows the mass, stiffness, and damping of the three modes and their orientations, which were calculated by this procedure.

Figure 5.9 compares the results of the identification method with the measurements of G_{zz}. The smooth curve is the receptance curve computed using the modal parameters of Table 5.2 and is superimposed on the experimental data. The figure shows good agreement between the two curves.

5.3 DYNAMICS OF THE CUTTING PROCESS

As mentioned in the introduction during orthogonal cutting, the cutting edge of the tool is perpendicular to the feed direction as shown in Figure 5.10. Although this configuration simplifies the problem, there are still a large

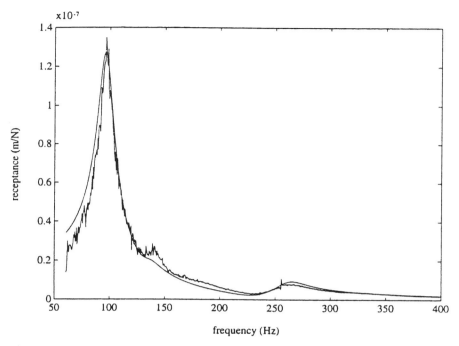

FIGURE 5.9. Computed versus measured receptance G_{zz}.

number of parameters that effect the cutting process. These parameters include work and tool material properties; tool geometrical characteristics such as rake, clearance and lead angles, and tool sharpness; nominal cutting conditions such as feed, depth of cut, and rotational and surface speed. An analytical cutting model should reflect the effect of all these parameters.

The quantities used in the formulation of the problem are shown in Figure 5.10. Note that the coordinate system of Figure 5.10 (i.e., the process coordi-

TABLE 5.2
Modal Parameters of the Major Structural Modes of the Hardinge Superslant AB Lathe

Mode	m_i(kg)	b_i(Ns/m)	k_i(N/m)	ϕ_i	ψ_i
$i = 1$	65.0	6.2×10^3	2.4×10^7	46°	6°
$i = 2$	135.9	17.1×10^3	9.8×10^7	70°	40°
$i = 3$	119.5	24.2×10^3	32.1×10^7	47°	32°

Source: Tembo (1994).

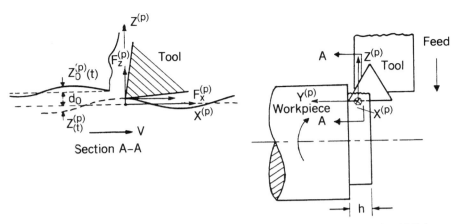

FIGURE 5.10. Geometry of orthogonal cutting. (From Minis et al., 1990b.)

nate system) is *different* from the coordinate system of the machine tool structure. The process coordinate system (p) moves along the nominal path of the tool; the axis $Z^{(p)}$ is normal to the work surface, and the axis $X^{(p)}$ has the direction of the surface speed. The important variables to be related, for a given cutting tool and workpiece, are the force applied on the tool and the resulting relative displacement of the cutting edge with respect to the workpiece. Wu and Liu (1985a, 1985b) used the minimum shear energy principle and assumed a small amplitude of oscillations to derive a set of nonlinear equations that relates the two components $F_x^{(p)}$ and $F_z^{(p)}$ of the cutting force vector to the $z^{(p)}$ component of the tool's displacement, which is also called the inner chip modulation. The Wu-Liu cutting force equations have been linearized around the solution of steady cutting in (Minis, 1988). The following relationships are the resulting unidirectional, linear, Fourier transformed cutting equations:

$$F_z^{(p)} = h(K_{di}z^{(p)} + K_{do}z_o^{(p)}),\qquad(10)$$

$$F_x^{(p)} = h(K_{ci}z^{(p)} + K_{co}z_o^{(p)}),\qquad(11)$$

where h is the depth of cut (width of the chip) and $z_o^{(p)}$ is the outer chip modulation, namely the surface displacement left by the tool on the workpiece during the previous revolution of the spindle. In the frequency domain

$$z_o^{(p)} = z^{(p)}e^{-j\omega T},\qquad(12)$$

where T is the period of revolution in turning (or the time between consecutive tooth engagements in milling). The complex constants K_{di}, K_{do}, K_{ci}, and K_{co}

of Equations (11) and (12) satisfy the following relationships:

$$K_{di} = \text{Re}(K_{di}) + j\,\text{Im}(K_{di}), \tag{13}$$

$$K_{do} \cong -\,\text{Re}(K_{di}) \tag{14}$$

$$K_{ci} = \text{Re}(K_{ci}) + j\,\text{Im}(K_{ci}), \tag{15}$$

$$K_{co} = -\,\text{Re}(K_{co}) - j\,\text{Im}(K_{co}). \tag{16}$$

The real parts, $\text{Re}(K_{di})$, $\text{Re}(K_{ci})$, are independent of the angular frequency ω and the imaginary parts, $\text{Im}(K_{di})$, $\text{Im}(K_{ci})$, are proportional to ω. According to these relationships only the determination of the inner transfer functions K_{di} and K_{ci} is required.

The determination of K_{di} and K_{ci} from force and displacement measurements requires the elimination of the influence of the outer chip modulation $z_o^{(p)}$. If $z_o^{(p)}$ is zero, then Equations (10) and (11) reduce to

$$F_z^{(p)} = hK_{di}z^{(p)}, \tag{17}$$

$$F_x^{(p)} = hK_{ci}z^{(p)}. \tag{18}$$

In this case also the intrinsic coupling of the machining system during cutting presents difficulties in the identification of the cutting process dynamics from input-output measurements. According to Åström and Eykhoff (1971) and Box and MacGregor (1974), such a process can be identified by introducing an external input (external force in the present case). This force should have the following characteristics: (1) It should be uncorrelated to the respective cutting force component; and (2) its magnitude should be comparable to the magnitude of the dynamic cutting force. Figure 5.11 illustrates the use of such a force in the identification method of Minis et al. (1990b). The figure shows the application of the method for the estimation of K_{di} using the experimental

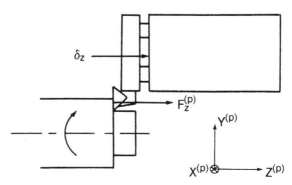

FIGURE 5.11. Schematic representation of the identification method for K_{di}. (From Minis et al., 1990b.)

apparatus of Figure 5.2. During the steady orthogonal turning of a solid steel workpiece (in which $z^{(p)} = 0$), an impulsive force δ_Z is applied by an impulse hammer directly to the lathe's turret in the $Z^{(p)}$ direction. The component $z^{(p)}$ of the structure's response at the tool tip is the inner modulation of the chip being cut. On the other hand, the outer chip modulation retains the zero value of steady cutting during the first workpiece revolution after the initial excitation. Hence the inner modulation is the only input to the cutting process and generates a dynamic cutting force that is measured by the dynamometer. Since the impulse is not applied on the dynamometer's cover plate, the load being measured is the cutting force, which is the output of the cutting process and not the total force, which includes the external excitation δ_Z. If both $z^{(p)}$ and $F_Z^{(p)}$ are known, then K_{di} can be determined from

$$K_{di} = \frac{F_z^{(p)}(\omega)}{hz^{(p)}(\omega)}, \tag{19}$$

where h is the known depth of cut. If a similar excitation δ_x is imposed in the $x^{(p)}$ direction and both $z^{(p)}$ and $F_x^{(p)}$ are measured, then K_{ci} can be obtained in a similar manner.

The identification of the cutting process by the method above is possible only in a very narrow bandwidth around the dominant peak of the structure's response, $z^{(p)}$. The values of K_{di} and K_{ci} were found at the frequency ω_0 that corresponds to the maximum value of the spectrum of the tool tip's displacement. Then the inner transfer functions can be evaluated for a range of frequencies from

$$\text{Re}[K_{ai}(j\omega)] = \text{Re}[K_{ai}(j\omega_0)] = \text{constant}, \qquad a = d, c, \tag{20}$$

$$\text{Im}[K_{ai}(j\omega)] = \text{Im}[K_{ai}(j\omega_0)] \frac{\omega}{\omega_0}. \tag{21}$$

Typical experimental results are presented in Figure 5.12, which shows the parameters $\text{Re}[K_{di}]$ and $\text{Im}[K_{di}]$ as a function of the surface speed for cutting mild steel 1018 with Kennametal TPMR 322 tools (5° rake angle and 4° clearance angle) and a feed rate of 0.076 mm/rev. All the experimental values shown were determined at $f_0 = \omega_0/2\pi = 107$ Hz. The graphs present three measurements of real and imaginary value pairs, each designated by a distinct symbol. These values were not averaged, since some parameters of the cutting system, such as the tool wear, varied slightly between tests. It is pointed out that according to the orientation of the cutting coordinate system (p), the values of the inner transfer functions are opposite to those indicated in the figures, where the convention of positive stiffness has been followed. The parameters $\text{Re}[K_{ci}]$ and $\text{Im}[K_{ci}]$ show similar trends (Minis et al., 1990b).

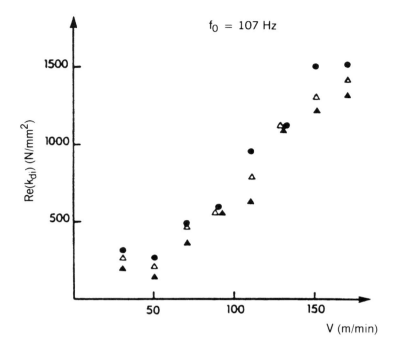

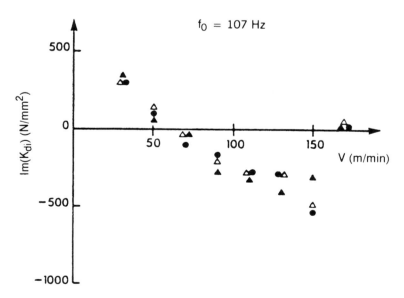

FIGURE 5.12. The values of $\mathrm{Re}[K_{di}]$ and $\mathrm{Im}[K_{di}]$ at $f_0 = 107\,\mathrm{Hz}$ for a feed rate of $0.076\,\mathrm{mm/rev}$. (From Minis et al., 1990b.)

5.4 PREDICTION OF MACHINE TOOL CHATTER

The structural model of Section 5.2.1 and the force relationships of Section 5.3 can be combined into a system of linear differential-difference equations that represent the dynamics of machining. The stability of the system's equilibrium point is related to machine tool chatter. To illustrate an important issue in forming the dynamical machining equations, consider the general turning configuration of Figure 5.13. In this case the cutting process' coordinate system (p) has a different orientation from the structure's coordinate system (m). Hence a coordinate transformation is needed to transform the cutting force vector from the process, (p)-system, to the machine, (m)-system; thus

$$\mathbf{F}^{(m)} = L\mathbf{F}^{(p)}, \tag{22}$$

where L is the transformation matrix. Conversely, the inverse transformation L^{-1} is required to transform the structure's response from the (m)-system back to the cutting process coordinate system (p); namely

$$\mathbf{r}^{(p)} = L^{-1}\mathbf{r}^{(m)}. \tag{23}$$

Using Equation (22), we see that

$$\mathbf{r}^{(p)} = L^{-1}GL\mathbf{F}^{(p)} = G'\mathbf{F}^{(p)}, \tag{24}$$

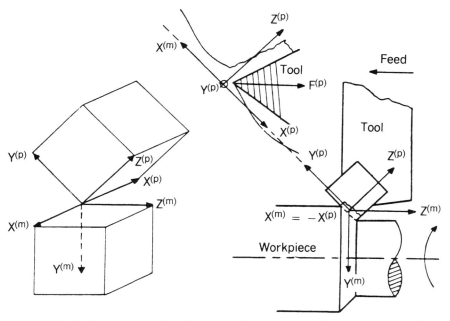

FIGURE 5.13. Relative orientation of the machine tool (m) and process (p) coordinate systems for a general turning configuration. (From Minis et al., 1990c.)

where G' is the transformed receptance matrix. The equation above clearly shows that the relative orientation of the two coordinate systems is critical to the system's stability, since it affects the magnitude of the elements of the receptance matrix.

5.4.1 Prediction of Chatter in Turning

The very simple geometry of right-handed turning is shown in Figure 5.14. In this case the normal to the surface of the cut, $z^{(p)}$, coincides with the direction $Z^{(m)}$ of the structure's coordinate system, and the work surface speed coincides with the $X^{(m)}$ direction. The orientations of the two systems are therefore identical, and the transformation matrix is the identity matrix. Under those conditions only the displacement component $z^{(m)} = z^{(p)}$ will affect the two cutting force components $F_x^{(p)}$ and $F_z^{(m)}$. Hence only the two elements G_{zz} and G_{zx} of the receptance matrix are needed to describe the dynamics of the structure in the stability analysis.

The characteristic equaton of right-handed turning is derived with reference to Figure 5.15. Inputs n_z and n_x represent the random components of the cutting force. After some straightforward algebraic manipulations the system's characteristic equation is obtained as

$$1 + h[-G_{zz}(K_{di} + e^{-j\omega T}K_{do}) - G_{zx}(K_{ci} + e^{-j\omega T}K_{co})] = 0. \qquad (25)$$

Equation (25) is a special case of the characteristic equation of turning dynamics. The general dynamical equations of turning assume the following form:

$$\mathbf{r}^{(m)} + hG(j\omega)LP(j\omega)L^{-1}\mathbf{r}^{(m)} = [I + hG(j\omega)A(j\omega)]\mathbf{r}^{(m)} = 0, \qquad (26)$$

where $\mathbf{r}^{(m)}$ is the tool displacement vector in the machine coordinate system; I is the 3×3 identity matrix; h is the depth of cut; L is the transformation matrix of Equation (22); $P(j\omega)$ is a 3×3 matrix representing the process dynamics, it relates the cutting force vector $\mathbf{F}^{(p)}$ with the tool displacement vector $\mathbf{r}^{p)}$ (see Equations (10) and (11)]; $G(j\omega)$ is the receptance matrix of the machine tool structure.

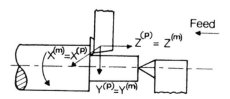

FIGURE 5.14. Geometry of right-handed turning. (From Minis et al., 1990c.)

Cutting process

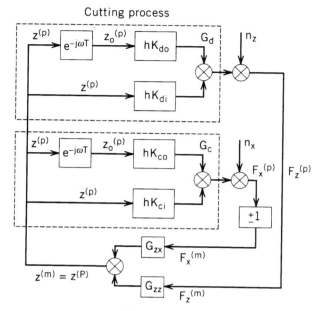

Machine Tool

FIGURE 5.15. Block diagram of orthogonal turning: $+1$ corresponds to right-handed turning, and -1 to left-handed turning. (From Minis et al., 1990c.)

Equations (25) and (26) suggest that the stability of turning depends on the following conditions: (1) the depth of cut, which can be thought of as the system gain; (2) the orientation of the tool with respect to the structure's coordinate system; (3) the spindle speed, which determines the period of revolution T; and (4) the surface speed v and the feed rate d_o, which, along with the workpiece material and the tool's geometry, affect the values of the cutting process transfer functions K_{di}, K_{do}, K_{ci}, and K_{co}.

The presence of the time delay terms in the characteristic equation (25) leads to the use of the Nyquist stability criterion in order to predict (1) the critical value of the depth of cut h_o, beyond which the machining dynamics are unstable and chatter develops, and (2) the chatter frequency. Consider the application of this stability analysis method for the Hardinge Superslant AB lathe. The machine tool receptances G_{zz} and G_{zx} of Section 5.2 are employed in Equation (25) along with the cutting process parameters K_{di}, and K_{ci} of Section 5.3. Figure 5.16 presents the theoretical predictions along with experimental measurements of h_c for right-handed turning of mild steel 1018 with the same tool as the one employed in the identification experiments of Section 5.3. The stability limit is shown as a function of the surface speed for a feed rate value of 0.076. Three predictions of h_c are given, each corresponding to one pair of the cutting parameters K_{di} and K_{ci} of Section 5.3. Figure 5.16 indicates good

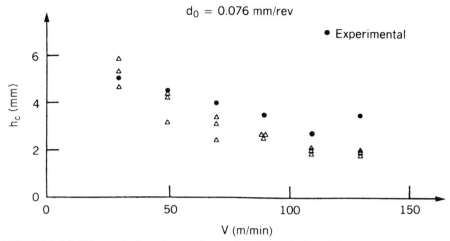

FIGURE 5.16. Theoretical and experimental values of the critical depth of cut for right-handed turning and a feed rate of 0.076 mm/rev. (From Minis et al., 1990c.)

agreement between the experimental measurements and the theoretical estimates of the stability limit. Figure 5.17 shows the variation of the critical depth of cut h_c with the surface and spindle speeds for right-handed turning and feed rate of 0.127 mm/rev. For each value of the surface speed, the figure shows the characteristic lobes of h_c in terms of the spindle speed. The quality of the predictions of Figure 5.16 is typical for various turning configurations, cutting tools and cutting conditions (see Minis et al., 1990c). This indicates that the linear approximation of the dynamics of machining and the linear stability analysis presented above are capable of predicting the critical depth of cut h_c.

An alternative way of chatter prediction is through simulation of the tool's motion using the modal Equations (3), (4), and (6)—and the counterparts of Equation (6) for $y^{(m)}$ and $z^{(m)}$. This way nonlinear models of the cutting process dynamics can be used in the stability analysis to account for significant process nonlinearities. For an example of such analysis, see Tembo (1994).

5.4.2   Prediction of Chatter in Milling

Consider the simple system of Figure 5.18 which performs half-immersion up-milling with a straight tooth cutter. In this case $X-Y$ is the coordinate system of the structural dynamics (system (m), whose axes are aligned with the principal modes of oscillation of the spindle. Note that the stiffness of the structure is considered infinite along the Z-axis (axis of the spindle), and thus two-dimensional coordinate systems are used. The coordinate system U_i-V_i corresponds to tooth i of the milling cutter and is rotating with the spindle speed Ω. With respect to this system tooth i may be viewed as a single point

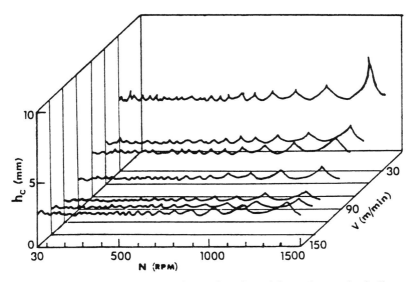

FIGURE 5.17. The critical depth of cut h_c as a function of the surface and spindle speeds for a feed rate of 0.127 mm/rev. (From Minis et al., 1990c.)

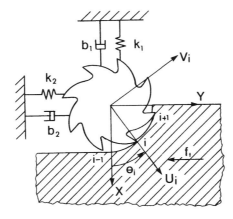

FIGURE 5.18. Two-degree-of-freedom milling system. (From Minis et al., 1990d.)

cutting tool in turning, and Equations (10) and (11) may be used to provide the cutting forces applied on this tooth. It is emphasized, however, that the time period T in the cutting force equations is the period between successive tooth engagements ($T = 2\pi/\mu\Omega$), where μ is the total number of teeth). The system U_i–V_i is analogous to the process system (p) of turning. Angle θ_i provides the relative orientation between the X–Y and U_i–V_i coordinate systems.

The dynamics of milling are described by an equation similar to Equation (26). In this case, however, the transformation matrix L_i between the structural and process systems is no longer constant but is a periodic function of time with period T. Superimposing the cutting forces applied to all teeth of the cutter, the dynamical equation for the tool motion is written in the time domain as follows (see Minis and Yanushevsky, 1993):

$$\mathbf{r} + hG(D) \sum_{i=0}^{\mu-1} L_i(t)P_i(D)L_i^{-1}(t)\mathbf{r} = [I + hG(D)A(D, t)]\mathbf{r} = 0, \qquad (27)$$

where $\mathbf{r} = (x \ y)^T$ is the tool displacement vector expressed in the machine coordinate system; D is the differential operator $d(\ldots)/dt$; I is the 2×2 identity matrix; h is the axial depth of cut; $L_i(t)$ is the periodic transformation matrix for tooth i; $P_i(D)$ is a 2×2 operator that represents the process dynamics at tooth i based on Equations (10) and (11) and accounts for the fact that, when the tooth is not in contact with the workpiece, the corresponding cutting force is 0; $G(D)$ is the receptance matrix of the machine tool structure in which $j\omega$ has been substituted with the differential operator D. Finally $A(D, t)$ is a 2×2 matrix operator that is periodic with period T.

Note that $\mathbf{r} = G(D)\mathbf{F}$. Substituting this expression into Equation (27) and multiplying by $G^{-1}(D)$ from the left, we obtain

$$\mathbf{F} + hA(D, t)G(D)\mathbf{F} = 0. \qquad (28)$$

This equation describes the dynamics of the milling system in terms of the cutting force $\mathbf{F}$, and it can be used to determine the critical values h_c of the axial depth of cut h for which the system becomes unstable. A straightforward but expensive way to do so is to integrate Equation (28) numerically for increasing values of h until chatter is detected (see Tlusty and Ismail, 1983). An alternative analytical method has been developed in Minis and Yanushevsky (1993). This method uses Floquet theory, Fourier analysis, and the properties of parametric transfer functions to derive the following approximation of the characteristic equation for the milling dynamics of Equation (28):

$$\det\left[I + hG(\lambda) \frac{1}{T} \int_0^T A(\lambda, t)dt\right] = 0. \qquad (29)$$

For stability, all the roots (eigenvalues) λ of Equation (29) must have negative real parts. Given the receptance matrix G of the milling machine and the process matrix A, a form of the Nyquist stability criterion can be used to estimate the critical value h_c of the axial depth of cut beyond which the milling system is unstable. Figure 5.19 compares the values of h_c provided by the numerical simulation of the milling dynamics (h_c vs. Ω) with the results of the analytical method ($h_c^{(0)}$ vs. Ω) for the system of Figure 5.18. The system

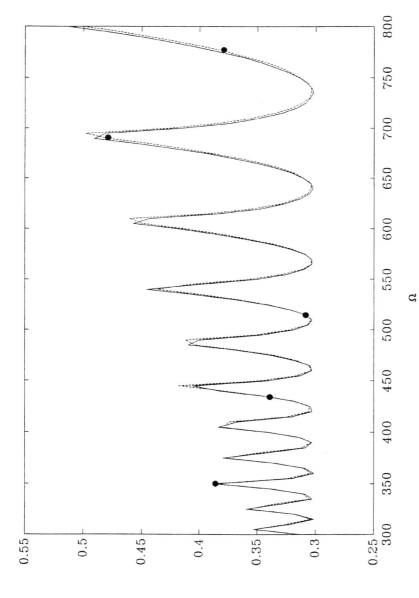

FIGURE 5.19. The critical value of the axial depth of cut versus spindle speed for the system of Figure 5.18: Solid line represents h_c obtained from numerical simulation; dashed line represents $h_c^{(o)}$ obtained from the analytical stability method. (From Minis et al., 1990d.)

parameters represent the practical milling case of reference (Tlusty and Ismail, 1983). The figure shows excellent agreement between the results of the two methods.

5.5   CHARACTERIZATION OF MACHINING DYNAMICS

Despite the intrinsic nonlinearity of machining dynamics, the discussion of Section 5.4 shows that linear models and techniques are appropriate for chatter prediction. This is not true, however, for other applications such as early chatter detection and control, which require a fundamental understanding of the dynamics involved. Below we discuss the use of basic and advanced signal-processing techniques for the characterization of machining dynamics from on-line measurements. Much of this discussion is motivated by recent advances in nonlinear dynamics, including the computation of fundamental dynamical measures in turning.

5.5.1   Power Spectra of Cutting Measurements

Figure 5.20a shows the power spectrum of the $X^{(m)}$ component of the tool acceleration during right-handed turning of steel 1018 workpieces. Turning is conducted with the Hardinge lathe equipped with a tool identical to the one discussed in Section 5.3. The cutting conditions were depth of cut $h = 2.8$ mm, spindle speed $\Omega = 307$ rpm (5.12 Hz), surface speed $v = 90$ m/min and feed rate $d_o = 0.1778$ mm/rev. The depth-of-cut value was lower than but very close to the critical value h_c. The measurement parameters were sampling rate = 4096 Hz, cutoff frequency = 1100 Hz, number of points in the FFT = 4096, and number of time records averaged $\cong 50$. The spectra exhibit a multiplicity of equally spaced peaks surrounding the first natural frequency of the lathe's structure (see also Figure 5.3). The differences between the frequency coordinates of the numbered peaks approximate the spindle speed within the resolution of the frequency scale. This is true for all acceleration and force measurements in over 200 experiments conducted within a wide range of cutting conditions (Chavali, 1995).

The sideband structure observed in Figure 5.20a is attributed to the inherent regeneration of the turning process which is modeled by the time delay of Equation (12). Note that time-delayed systems exhibit an infinite number of distinct poles, which are manifested as distinct peaks in the spectrum. This observation is supported by Figure 5.20b which shows the power spectrum of the tool acceleration along the $X^{(m)}$-axis measured during a thread-cutting experiment. During thread cutting there is no overlap of successive cuts and thus no regeneration. In this case the spectrum does not exhibit the sideband structure of Figure 5.20b. Only one peak is seen about the second natural mode of the machine tool structure. Note that the second mode is excited because thread cutting is not orthogonal and induces a strong cutting force component in the radial $Y^{(m)}$ direction.

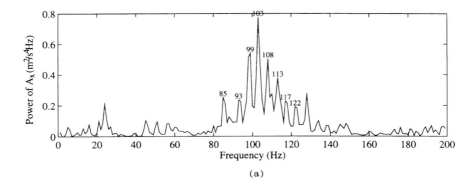

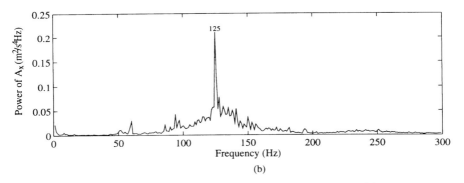

FIGURE 5.20. Power spectra of the tool acceleration in the $X^{(m)}$ direction: (a) Orthogonal cutting with regeneration. (b) Thread cutting with no regeneration.

5.5.2 Delay Space

The estimation of attractor dimension, Lyapunov exponents, false nearest neighbors, and other system invariants from a given scalar time series, $s(t)$, requires as a first step the construction of a state space for the underlying dynamical system. The use of delays in constructing a state space was suggested by (Ruelle, 1979; Packard, 1980) and may be described as follows:

Given $s(t)$, a point in the R^n delay space, $X(t) = \{x_0(t), x_1(t), \ldots, x_{n-1}(t)\}$, has components $x_k(t) = s(t + k\tau)$, where τ is the time delay. The results of Mane (1981) and Takens (1981), which are predicated on noise-free time series, indicate that the choice of the delay τ is arbitrary and that for n sufficiently large properties of the system's attractor based on a delay space reconstruction will be identical to those of an attractor constructed from the physical coordinates required to describe the system. A recent discussion of the topic is presented in Sauer (1991).

It is evident from the definition of time delay coordinates that for small τ coordinates $x_m(t)$ and $x_{m-1}(t)$ could become indistinguishable. The problem is exacerbated by numerical and experimental truncation and roundoff but more seriously by experimental noise. If, on the other hand, the delay τ is too large, then $x_m(t)$ and $x_{m+1}(t)$ may be statistically completely unrelated. An effective algorithm for the determination of τ based on mutual information was given in Fraser and Swinney (1986).

5.5.3 Mutual Information

Fraser and Swinney suggested that τ be chosen as the first local minimum, when it exits, of the mutual information function $I(x_0; x_1)$ computed for $x_0(t)$ and $x_1(t)$:

$$I(X; Y) = \int P_{XY}(x, y)\log_2 \left[\frac{P_{XY}(x, y)}{P_X(x)P_Y(y)} \right] dxdy, \tag{30}$$

where P_{XY} is the joint probability distribution, and P_X and P_Y are probability distribution functions. If x and y are completely independent, then the argument of $\log_2$ is unity and $I(X; Y) = 0$. The probability distribution functions were evaluated through histograms that proved to be effective for time series, often computer generated, with high signal-to-noise ratios. However, histogram evaluations proved to be inadequate for the noisy time series derived from cutting meausrments (Berger and Minis, 1995).

In a code for the evaluation of mutual information, Kennel (1992a), evaluated probability distribution functions with smoothing and adaptive kernel density estimators rather than histograms. Kennel's code proved to be highly effective in the calculation of mutual information by time series derived from cutting data.

Mutual information was computed for 25 sets of cutting tool acceleration data utilizing Kennel's algorithm. The data were measured during right-handed turning with the apparatus of Sections 5.2 and 5.3. These results are summarized in Berger and Minis (1995). A typical result is provided by the analysis of pre-chatter data set 201 (tool acceleration along the $X^{(m)}$ direction recorded during cutting mild steel 1018 with $h = 2.0$ mm, $d_0 = 0.007$ mm/rev, $v = 90$ m/min, $\Omega = 430$ rpm) retaining 50,000 data points in the calculation. A well-defined minimum is seen to occur at $\tau = 10$ and a maximum at $\tau = 21$; Figure 5.21. The first natural frequency of the cutting system $\cong 97$ Hz (see also Figure 5.3). Since the data sampling rate is 4096 Hz, it follows that $\tau = 10$ corresponds to one-quarter of the first natural frequency. The mutual information algorithm, utilizing smoothing and adaptive kernel density estimators, is sufficiently robust to determine the first minimum of $I(x_0, x_1)$ for all of the 25 data sets studied.

5.5.4 Autocorrelation Functions

The autocorrelation function, $\rho(k)$ is defined as (Box and Jenkins, 1994)

$$\rho(k) = \frac{E[(s(t) - \mu)(s(t + k) - \mu)]}{\sqrt{E[(s(t) - \mu)^2]E[(s(t + k) - \mu)^2]}}, \tag{31}$$

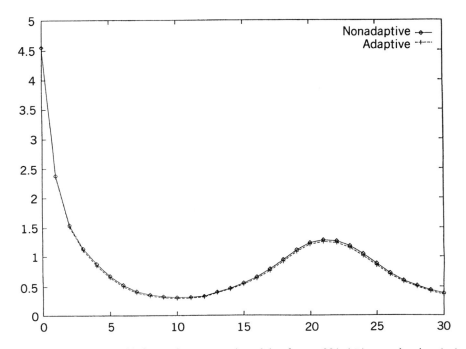

FIGURE 5.21. Mutual information versus time delay for set 201: (○) nonadaptive; (+) adaptive.

where $E[s(t)]$ is the expected value of the time series, $s(t)$, and the mean, μ, can be estimated by the sample mean

$$\mu = \frac{1}{N} \sum_{n=1}^{N} s(n). \tag{32}$$

The autocorrelation function measures the linear dependence of a function, and the first zero of the autocorrelation function approximates the first minimum of the mutual information. Figures 5.22 and 5.23 show the first zeros of the autocorrelation function for data sets 101 and 201 containing 163,840 points. (Data set 101 represents also tool acceleration along the $X^{(m)}$ direction recorded during cutting mild steel 1018, with $h = 1.0 \, mm$, $d_0 = 0.007 \, mm/rev$, $v = 90 \, m/min$, and $\Omega = 430 \, rpm$.) The first zeros are seen to occur at $k = 8.0$ and 9.5, which corresponds closely to the optimum delays of $\tau = 8.0$ and 10.0 given by the mutual information algorithm.

5.5.5 Singular Values

The autocovariance at lag k is defined by

$$\Gamma_k = E[(s(t) - \mu)(s(t+k) - \mu)]. \tag{33}$$

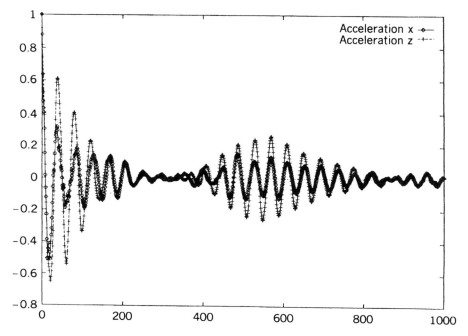

FIGURE 5.22. Autocorrelation function versus lag for set 101.

The covariance matrix

$$
\Gamma_n = \begin{pmatrix}
\gamma_0 & \gamma_1 & \gamma_2 & \cdots & \gamma_{n-1} \\
\gamma_1 & \gamma_0 & \gamma_1 & \cdots & \gamma_{n-2} \\
\gamma_2 & \gamma_1 & \gamma_0 & \cdots & \gamma_{n-3} \\
\cdots & & & \cdots & \cdots \\
\gamma_{n-1} & \gamma_{n-2} & \gamma_{n-3} & \cdots & \gamma_0
\end{pmatrix}
\tag{34}
$$

is symmetric and Toeplitz (Box et al., 1994). In Broomhead and King (1986) global singular values and singular vectors of the covariance matrix, (34), were used in the analysis and attractor reconstruction of nonlinear sytems. In Albano et al. (1988), Vantard and Ghil (1989), and Fraedrich and Wang (1993) the global singular value methods of Broomhead and King (1986) were applied to the calculation of attractor dimension. Limitations of the method have been noted by several authors, most recently Palus and Dvorak (1992). For linear systems the method gives linearly independent singular vectors and corresponding linearly independent coordinates. However, the method may fail to give independent coordinates, in the nonlinear sense, for nonlinear systems. In either case the set of singular vectors is complete.

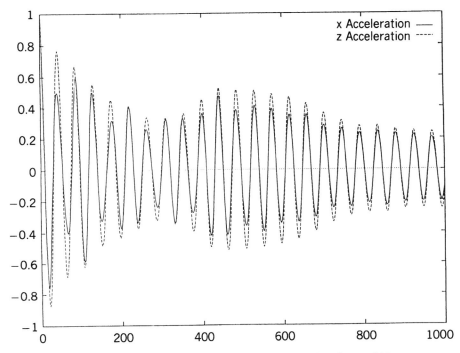

FIGURE 5.23. Autocorrelation function versus lag for set 201.

Given a real $m \times n$ matrix A, then there exists orthogonal matrices $U = [u_1, \ldots, u_m] \varepsilon \mathbf{R}^{m \times m}$ and $V = [v_1, \ldots, v_n] \varepsilon \mathbf{R}^{n \times n}$ such that

$$U^T A V = \text{diag}(\sigma_1, \ldots, \sigma_p) \varepsilon \mathbf{R}^{m \times n}, \qquad (35)$$

where $p \equiv \min\{m, n\}$ and $\sigma_1 \geqslant \sigma_2 \geqslant \cdots \geqslant \sigma_p \geqslant 0$ (Golub and Van Loan, 1993). Singular values σ_i of the covariance matrix Γ_n, Equation (34), were computed for twenty-five sets of cutting data. Typical results, provided by data sets 261 and 262 (tool accelerations along the $X^{(m)}$ and $Z^{(m)}$ directions, respectively, recorded during cutting mild steel 1018 with $d_0 = 0.007$ mm/rev, $\Omega = 750$ rpm, $v = 90$ m/min, and $h = 2.6$ mm) are shown in Figures 5.24 and 5.25. The first eight singular vectors for data set 261 are shown in Figure 5.26. The orthogonality of these has been verified by direct calculation. Projections of the time series onto the eight individual singular vectors are shown in Figure 5.27. The independence of the first two projections has been verified through calculation of the mutual information between the two (Fraser and Swinney, 1986, Kennel, 1992a). The mutual information was found to have a minimum of small magnitude for $\tau = 0$ and increases monotonically with τ. These two

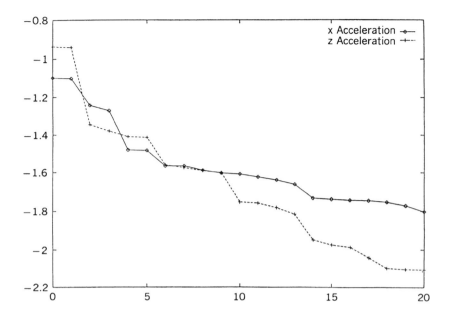

FIGURE 5.24. Log 10 of singular values versus lag for set 261.

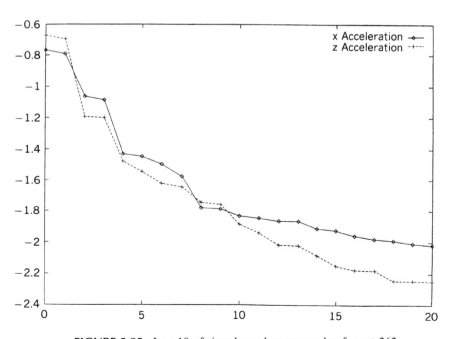

FIGURE 5.25. Log 10 of singular values versus lag for set 262.

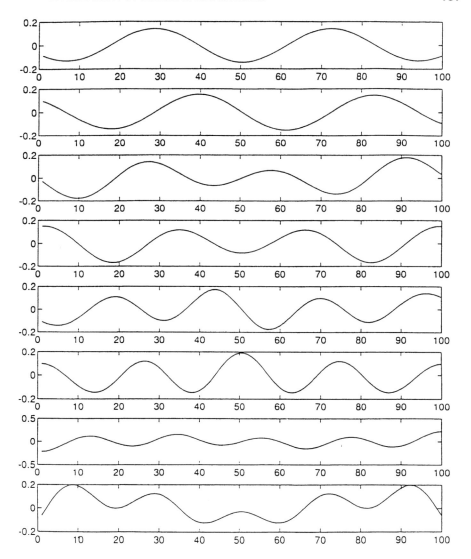

FIGURE 5.26. First eight singular vectors for set 261.

projections, x_1 and x_2, constitute a representation of the projection of the system's attractor onto the x_1x_2 plane. As in the case of data set 261, the mutual information between x_1 and x_2 had a minimum of small magnitude for $\tau = 0$ and increased monotonically with τ. The singular value reconstruction appears to suppress noise.

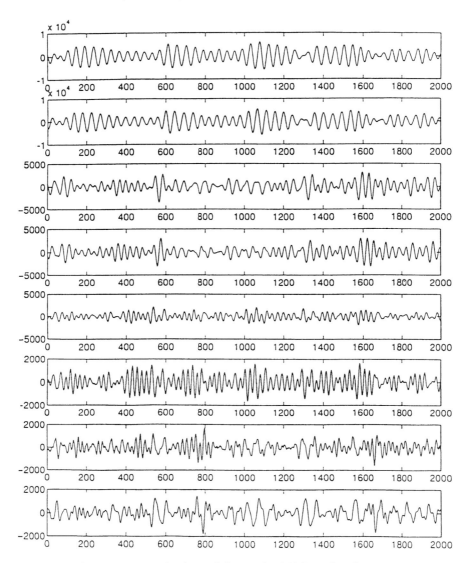

FIGURE 5.27. Projections of time series 261 into singular vectors.

5.5.6 Global False Nearest Neighbors

If the dimension n of the delay space reconstruction is a sufficient embedding dimension d_s, then the associated attractor is completely unfolded. d_s provides an upper bound on the attractor's dimension d_A. It is known that $d_s = 2d_A + 1$ (Whitney, 1937). The minimum embedding dimension may be less than d_s.

By tracking the behavior of near neighbors under unit increases in the dimension, d, of a delay reconstruction, the method of global false nearest neighbors, *fnn* (Kennel, et al., 1992b), provides an acceptable minimum delay embedding dimension d_E. Denote the rth nearest neighbor of $X(n)$ by $X^{(r)}(n)$ in R^d. The square of the Euclidean distance, $R_d(n, r)$, between $X(n)$ and $X^{(r)}(n)$ is

$$R_d^2(n, r) = \sum_{k=0}^{d-1} [s(n + k\tau) - s^{(r)}(n + k\tau)]^2 \tag{36}$$

Any neighbor for which

$$\left[\frac{R_{d+1}^2(n, r) - R_d^2(n, r)}{R_d^2(n, r)} \right]^{1/2} > R_{\text{TOL}} \tag{37}$$

is defined as a false neighbor where R_{TOL} is a threshold. R_{TOL} was assumed to be $\geqslant 10$. It was found that $d_E = 4$ for the twenty-five data sets studied by Berger and Minis (1995). A typical result is shown in Figure 5.28 for data set 202 which is in a pre-chatter state (tool acceleration along the $Z^{(m)}$ direction recorded during cutting mild steel 1018 with $d_0 = 0.007$ mm/rev, $\Omega = 430$ rpm,

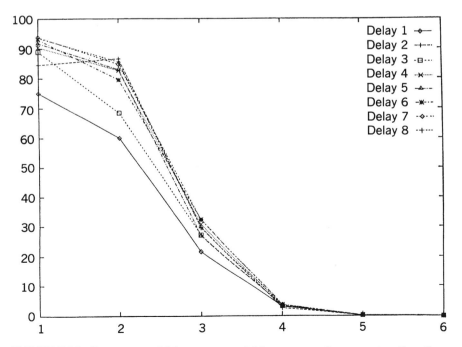

FIGURE 5.28. Percentage of false nearest neighbors versus the reconstruction dimension. (From Berger et al., 1995.)

$v = 90 \, \text{m/min}$, and $h = 2.0 \, \text{mm}$). A reduction in slope followed by a plateau at $d \geqslant 4$ indicates an acceptable value of $d_E = 4$. Since $d_A \leqslant d_E$, it follows that 4 is an upper bound on d_A. In this instance the function percentage of false nearest neighbors versus reconstruction dimension is nearly independent of delay, $1 \leqslant \tau \leqslant 11$. A delay $\tau = 11$ was the optimum provided by the mutual information calculation (30).

5.6 CONCLUSIONS

This chapter has presented methods to model and analyze the dynamics of machining for the prediction of machine tool chatter. The intrinsic coupling of the structural dynamics with the process dynamics during cutting was modeled by a closed-loop system. This coupling provides certain difficulties in identification. Two alternative experimental methods were outlined to determine the receptance matrix of machine tool structures as well as their most significant structural modes. In addition the chapter has provided a simple method for the identification of the parameters of a set of linearized cutting force equations. The structural and process models were used to predict the critical depth of cut beyond which chatter develops. For this purpose a unified stability method was discussed and applied to turning and milling. Comparison of the analytical predictions with experimental results showed that the stability method is capable of predicting the onset of chatter in general cutting configurations and under a variety of cutting conditions. Thus linear models and techniques are appropriate for chatter prediction.

More advanced applications, however, such as early chatter detection and control require a fundamental understanding of the dynamics involved. The chapter discussed the use of basic and advanced signal-processing techniques for the characterization of machining dynamics from on-line measurements. Specifically it was shown that certain characteristics of the power spectrum of cutting measurements justify the use of time delays in machining models. Furthermore, the computation of the optimal delay as the first minimum of the mutual information, $I(x_0, x_1)$, was achieved through an algorithm based on smoothing and adaptive kernel density estimators. For the data sets studied, the first minimum of the mutual information was well approximated by the first zero of the autocorrelation function. It was found that the mutual information has a small magnitude for zero delay and increases monotonically with delay for projections of the time series onto the singular vectors associated with the first two singular values. A two-dimensional singular vector projection of the attractor indicates that singular value reconstruction suppresses noise. Finally it was found that the upper bound of the attractor's dimension for right-handed orthogonal turning is 4; that is, the system dynamics can be described with at most four first-order differential equations.

REFERENCES

Ahn, T. Y., Eman, K. F., and Wu, S. M. 1985. Cutting dynamics identification by dynamic data system modeling approach. *ASME J. Eng. Ind.* **107**:91–94.

Albano, A. M., Muench, J., Schwrtz, C., Mees, A. I., and Rapp, P. E. 1988. Singular value decomposition and the Grassberger-Procaccia algorithm. *Phys. Rev. A* **38**:3017–3026.

Astrom, K. J., and Eykhoff, P. 1971. System identification—A survey. *Automa.* **7**.

Berger, B. S., Minis, I., Chen, Y. H., Chavali, A., and Rokni, M. 1995. Attractor embedding in metal cutting. *J. Sound Vibr.* **184**:936–942.

Box, G. G. P., and MacGregor, J. F. 1974. The analysis of closed loop dynamic stochastic systems. *Technomet.* **16**.

Box, G. G. P., Jenkins, G. M., and Reinsel, G. C. 1994. *Time Series Analysis*. Prentice Hall, Englewood Cliffs, NJ.

Broomhead, D. S., and King, G. P. 1986. Extracting qualitative dynamics from experimental data. *Phys. D*, **20**:217–236.

Chavali, A. 1995. Analysis of time series in metal cutting dynamics. M.S. thesis. Department of Mechanical Engineering., University of Maryland, College Park.

Delio, T., Tlusty, J., and Smith, S. 1992. Use of audio signals for chatter detection and control. *ASME J. Eng. Ind.* **114**:146–157.

Eman, K., and Wu, S. M. 1980. A feasibility study of on-line identification of chatter in turning operations. *ASME J. Eng. Ind.* **102**:315–321.

Ewins, D. J. 1984. *Modal Testing: Theory and Practice*. Research Studies Press/Wiley, New York.

Fisher, E. E. 1965. *Identification of Linear Systems*. Reprints JACC, Troy, NY.

Fraser, A. M., and Swinney, H. L. 1986. Independent coordinates for strange attractors from mutual information. *Phys. Rev. A* **33**:1134–1140.

Golub, G. H., and Van Loan, C. F. 1991. *Matrix Computations*. Johns Hopkins University Press, Baltimore, MD.

Grabec, I. 1986. Chaos generated by the cutting process. *Phys. Lett. A* **117**:384–386.

Grabec, I. 1988. Exploration of random vibrations in cutting on grounds of deterministic chaos. *Roboti. Comp.-Integrated Manuf.* **4**:129.–134.

Gustavsson, I., Ljung, L., and Sodestrom, T. 1977. Identification of process in closed loop. Identifiability and accuracy aspects. *Automat.* **13**.

Hanna, N. H. and Tobias, S. A. 1969. The non-linear dynamic behaviour of a machine structure. *Int. J. Mach. Tool Des. Res.* **9**:293–307.

Jemielniak, K., and Widota, A. 1989. Numerical simulation of chatter vibration in turning. *Int. J. Mach. Tools Manuf.* **29**:239–248.

Kennel, M. B. 1992a. The multiple mutual information program. Report. Institute for Nonlinear Science, University of California, San Diego, CA.

Kennell, M. B., Brown, R., and Abarbanel, H. D. I. 1992b. Determining embedding dimension from phase-space reconstruction using a geometrical construction. *Phys. Rev. A* **45**:6.

Kim, K. J., Eman, E. F., and Wu, S. M. 1984. Identification of natural frequencies and damping ratios of machine tool structures by the dynamic data system approach. *Int. J. Mach. Tool Des. Res.* **24**:161–170.

Klamecki, B. E. 1989. On the effects of process asymmetry on process dynamics. *ASME J. Eng. Ind.* **111**:193–198.

Koenigsberger, I., and Tlusty, J. 1971. *Structures of Machine Tools.* Pergamon Press, Manchester.

Lin, J. S., and Weng, C. I. 1990. A nonlinear dynamic model of cutting. *Int. J. Mach. Tools Manuf.* **30**:53–64.

Mane, R. 1981. In Rand, D., and Young, L. S., eds., *Dynamical Systems and Turbulence.* Lecture Notes in Math., No. 898. Springer-Verlag, New York, p. 230.

Merchant, M. E. 1945. Mechanics of the metal cutting process. I: Orthogonal cutting and a type 2 chip. *J. Appl. Phys.* **16**:267–275.

Merritt, H. E. 1965. Theory of self-excited machine tool chatter. *ASME J. Eng. Ind.*, **87**:447–454.

Minis, I. 1988. Prediction of machine tool chatter in turning. Ph.D. dissertation. Department of Mechanical Engineering, University of Maryland, College Park.

Minis, I., and Yanushevsky, R. 1993. A new theoretical approach for the prediction of machine tool chatter in milling. *ASME J. Eng. Ind.* **115**:1–8.

Minis, I., Magrab, E., and Pandelidis, I. 1990a. Improved methods for the prediction of chatter in turning. I: Determination of the structural response parameters. *ASME J. Eng. Ind.* **112**:12–20.

Minis, I., Magrab, E., and Pandelidis, I. 1990b. Improved methods for the prediction of chatter in turning. II: Determination of cutting process parameters. *ASME J. Eng. Ind.* **112**:21–27.

Minis, I., Magrab, E., and Pandelidis, I. 1990c. Improved methods for the prediction of chatter in turning. III: A generalized linear theory. *ASME J. Eng. Ind.* **112**:28–35.

Minis, I., Yanushevsky, R., and Tembo, A. 1990d. Analysis of linear and nonlinear chatter in milling. *Ann. CIRP* **39**:459–462.

Moriwaki, T., and Iwata, K. 1976. In-process analysis of machine tool structure dynamics and prediction of machine tool chatter. *ASME J. Eng. Ind.* **98**.

Packard, N. J., Crutchfield, J. P., Fromer, J. D., and Shaw, R. S. 1980. Geometry from a time series. *Phys. Res. Ltrs.* **115**:712–716.

Peters, J., Vanherck, P., and Van Brussel, H. 1971. The measurement of the dynamic cutting coefficient. *CIRP Ann.* **20**:129–136.

Palus, M., and Dvorak, I. 1992. Singular value decomposition in attractor reconstruction: Pitfalls and precautions. *Physica D* **55**:221–234.

Rahman, M. 1988. In-process detection of chatter threshold. *ASME J. Eng. Ind.* **110**:44–50.

Rahman, M., and Ito, Y. 1986. Detection of the onset of chatter vibration. *J. Sound Vib.* **109**:193–205.

Ruelle, D. 1979. Ergodic theory of differentiable dynamical systems. *Math. Inst. Hautes Etudes Sci.*, **5**:27.

Sauer, T., Yorke, J. A., and Casdagi, M. 1991. Embedology, *J. Stat. Phys.* **65**:579–616.

Shi, H. M., and Tobias, S. A. 1984. Theory of finite amplitude machine tool instability. *Int. J. Mach. Tools Manuf.* **24**:45–69.

Takens, F. 1981. In Rand, D., and Young, L. S., eds., *Dynamical Systems and Turbulence.* Lecture Notes in Math., No. 898. Springer-Verlag, New York, pp. 366–381.

Taylor, H. R. 1977. A comparison of methods for measuring the frequency response of mechanical structures with particular reference to machine tools. *Proc. Inst. Mech. Eng.* **191**:257–270.

Tembo, A. 1994. Analysis, prediction and control of machining dynamics applied to turning processes. Ph.D. dissertation. Department of Mechanical Engineering, University of Maryland, Baltimore.

Tlusty, J. 1978. Analysis of the state of research in cutting dynamics. *CIRP Ann.* **27**:583–589.

Tlusty, J., and Ismail, F. 1983. Special aspects of chatter in milling. *ASME J. Vib., Stress, Reliab. Des.* 105:24–32.

Tobias, S. A. 1965. *Machine Tool Vibration.* Wiley, New York.

Vantard, R., and Ghil, M. 1989. Singular spectrum analysis in nonlinear dynamics with applications to paleoclimatic time series. *Physica D* **35**: 394–424.

Weck, M. 1985. *Handbook of Machine Tools,* vol. 4. Wiley, New York.

Whitney, H. 1937. The imbedding of manifolds in families of analytical manifolds. *Ann. Math.* **37**:815–878.

Wu, D. W. 1988. Comprehensive dynamic cutting force model and its application to wave-removing processes. *ASME J. Eng. Ind.* **110**:153–161.

Wu, D. W. 1989. A new approach to formulating the transfer function for dynamic cutting processes. *ASME J. Eng. Ind.* **111**:37–47.

Wu, D. W., and Liu, C. R. 1985a. An analytical model of cutting dynamics. I: Model building. *ASME J. Eng. Ind.* **107**:107–111.

Wu, D. W., and Liu, C. R. 1985b. An analytical model of cutting dynamics. II. *ASME J. Eng. Ind.* **107**:112–118.

6

DELAY-DIFFERENTIAL
EQUATION MODELS
FOR MACHINE TOOL
CHATTER

GÁBOR STÉPÁN

6.1 INTRODUCTION

Machine tool chatter is one of the most complex dynamical processes. Since the accuracy of a machine tool is strongly affected by the vibrations arising during the cutting process, several models have appeared in the specialist literature to explain and to predict these vibrations. From dynamical systems viewpoint, the most complicated models are the ones that describe the so-called self-excited vibrations of the machine tools. Within this group of models, the complexity of the model increases with the number of degrees of freedom, namely with the dimension of the phase space where the trajectories are embedded. The greater the dimension of the phase space is, the more complex the described dynamical phenomena can be.

An essential cause of the vibrations in the cutting process is the regenerative effect. Its mechanical model can still be a single degree of freedom system, but the corresponding mathematical model is an infinite-dimensional one. The presence of the time delay results delay-differential equations (DDE), and the trajectories can uniquely be described in an infinite-dimensional phase space only. Since the mechanical model can still be simple and deterministic, there is a possibility for analytical, often closed-form calculations and the topological

Dynamics and Chaos in Manufacturing Processes, Edited by Francis C. Moon.
ISBN 0-471-15293-5 © 1998 John Wiley & Sons, Inc.

description of trajectories may present complicated, even chaotic nonlinear vibrations. If we take a careful look at the experimental results, they always contain more or less stochastic components. Since the numerical simulation of the complex models may or may not include stochasticity which refers to the dynamics of the model and the discretization technique together, it is important to predict nonlinear vibrations also with analytical work now involving computer algebra, too.

First, we summarize briefly the stability of linear autonomous DDEs. Some of the stability charts relevant in modeling oscillatory systems with delay are also presented. Then simple mechanical models are given for regenerative machine tool chatter. The construction of basic stability charts in the plane of the technological parameters is explained. Finally local and global nonlinear phenomena are described when the nonlinearity occurring in the DDE model is related to the cutting force. Delay differential equations with time periodic coefficients are not discussed here. Note, however, that such mathematical models are also important in modeling machine tool vibrations in case of milling, when the number of cutting edges working together varies in time (e.g., Minis, 1994).

6.2   GUIDE TO DELAY-DIFFERENTIAL EQUATIONS

The simplest DDE has the form

$$\dot{x}(t) = x(t-1), \tag{1}$$

where the state variable x is scalar, $x(t) \in \mathbb{R}$, the dot stands for differentiation with respect to the time t, and the time delay is just 1. The DDE describes a system where the present rate of change of state depends on a past value of the state. Substituting the trial solution $x(t) = Ke^{\lambda t}$, $K, \lambda \in \mathbb{C}$, we obtain a nontrivial solution for K when

$$(\lambda - e^{-\lambda})Ke^{\lambda t} \Rightarrow \lambda - e^{-\lambda} = 0.$$

The equation above is also called the characteristic equation, and it has infinitely many solutions for the complex characteristic roots $\lambda_j, j = 1, 2, \ldots$. As this simple example suggests, the theory of DDEs is a quite direct generalization of the theory of ordinary differential equations (ODE) into infinite-dimensional phase spaces. This generalization is not a trivial task, though, and it uses the mathematical tools developed for functional differential equations (FDE). For a thorough introduction to this theory, see Hale (1977) and Kuang (1993).

In case of linear time-independent mechanical models describing vibratory systems in the presence of time delay, the most general mathematical model assumes the form

$$\mathbf{M}\ddot{\mathbf{x}}(t) + \int_{-r}^{0} d\mathbf{B}(\theta)\dot{\mathbf{x}}(t + \theta) + \int_{-r}^{0} d\mathbf{C}(\theta)\mathbf{x}(t + \theta) = 0. \tag{2}$$

It has a similar structure to the well-known matrix differential equation of small oscillations in a finite-degree-of-freedom (DOF) system about its stable equilibrium. In this equation $\mathbf{x}(t) \in \mathbb{R}^n$, where the mechanical system has n DOF. The constant matrix $\mathbf{M}$ is the usual symmetric and positive definite mass matrix, while the matrices $\mathbf{B}, \mathbf{C}$ describe the weights of some past effects with respect to "damping" and "stiffness" in the system back in time until $t - r$. The elements of $\mathbf{B}(\theta) = [b_{jk}(\theta)]$, $\mathbf{C}(\theta) = [c_{jk}(\theta)]$ are functions of bounded variation, and the corresponding terms in (2) contain the so-called Stieltjes integrals. This kind of shorthand is very convenient in describing two different types of delays that also appear in regenerative machine tool chatter models. One is the *discrete delay* τ; the other is the *continuous delay* described by a weight function w over a certain interval. For example, the scalar

$$c(\theta) = \begin{cases} 0, & \theta \in [-r, -\tau), \\ -1, & \theta \in [-\tau, -h), \\ h - 1 + \theta + \dfrac{h}{\pi}\sin\left(\dfrac{\theta\pi}{h}\right), & \theta \in [-h, 0), \\ h, & \theta = 0, \end{cases} \tag{3}$$

in Figure 6.1 gives

$$\int_{-r}^{0} x(t + \theta)dc(\theta) = x(t) + \int_{-h}^{0}\left(1 + \cos\left(\dfrac{\theta\pi}{h}\right)\right)x(t + \theta)d\theta - x(t - \tau),$$

since $dc(\theta) = \dot{c}(\theta)d\theta$ where c is differentiable and since $dc(\theta) = c(\theta + 0) - c(\theta - 0)$ where c has a discontinuity. Thus the coefficients of the discrete delay terms at 0 and $-\tau$ are

$$c(+0) - c(-0) = +1 \quad \text{and} \quad c(-\tau + 0) - c(-\tau - 0) = -1,$$

while the weight function for the continuous time delay term is

$$w(\theta) = \dfrac{d}{d\theta}c(\theta) = 1 + \cos\left(\dfrac{\theta\pi}{h}\right) \qquad \theta \in [-h, 0).$$

The trivial solution $\mathbf{x} \equiv \mathbf{O}$ of the DDE (2) is not necessarily stable, of course. However, the necessary and sufficient condition for asymptotic stability is the same as it is for ODEs: The real parts of all the (infinitely many) characteristic

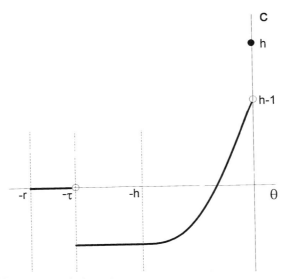

FIGURE 6.1. Description of discrete and continuous delays with (3).

roots are negative. These characteristic roots are the zeros of the transcendental characteristic function

$$D(\lambda) = \det \left(\mathbf{M}\lambda^2 + \int_{-r}^{0} \lambda e^{\lambda\theta} d\mathbf{B}(\theta) + \int_{-r}^{0} e^{\lambda\theta} d\mathbf{C}(\theta) \right) \qquad (4)$$

obtained by substituting the trial solution $\mathbf{K}e^{\lambda t}$ into the DDE (2), as shown in the introductory scalar example (1).

The infinite-dimensional version of the Routh-Hurwitz criterion used for the characteristic polynomial of ODEs is needed to analyze the zeros of the characteristic functions (4) of DDEs. If we define the real functions

$$R(\omega) = \operatorname{Re} D(i\omega), \quad S(\omega) = \operatorname{Im} D(i\omega), \qquad \omega \in [0, \infty), \qquad (5)$$

and denote the real zeros of R by $\rho_1 \geqslant \ldots \rho_r \geqslant 0$, that is,

$$R(\rho_k) = 0, \qquad k = 1, \ldots, r,$$

then the equilibrium of the linear delayed mechanical system (2) of n DOF is asymptotically stable if and only if

$$S(\rho_k) \neq 0, \qquad k = 1, \ldots, r, \qquad (6)$$

$$\sum_{k=1}^{r} (-1)^k \operatorname{sgn} S(\rho_k) = (-1)^n n. \qquad (7)$$

In n DOF systems the dimension of the phase space is $2n$, meaning always even. The above-mentioned stability criterion has a somewhat more complicated form for DDEs with odd dimensions as in (1) (see Stépán, 1989, for further discussion).

In the following section the most important stability charts are summarized for delayed vibratory systems. These charts are explained for simple cases having to do with the minimum number of mathematical parameters. Later these charts will be referred to when the equations of regenerative machine tool chatter are fully analyzed with the necessary mechanical and technological parameters and parameter functions.

6.3 BASIC STABILITY CHARTS OF DELAYED VIBRATORY SYSTEMS

A stability chart presents the domains of the system parameters where the equilibrium is asymptotically stable. By presenting the number of the pure imaginary (or zero) characteristic roots along the stability limits, it also refers to the way the equilibrium loses its stability. The stability limits can be determined in the parameter space (often a plane) by plotting the co-dimension 1 surfaces (often curves) given by the so-called D-curves

$$R(\omega) = 0, \quad S(\omega) = 0, \qquad \omega \in [0, \infty), \tag{8}$$

where R, S are defined as in (5) by means of the characteristic function D and ω takes the role of the parameter of the curve. In general, these curves separate infinitely many disjunct domains, and we need the stability criteria (6) and (7) to select the domains that correspond to asymptotic stability.

As an introductory example consider the simple first-order scalar DDE,

$$\dot{x}(t) + c \int_{-\infty}^{0} w(\theta)x(t + \theta)d\theta = 0, \qquad c \in \mathbb{R}. \tag{9}$$

The weight function

$$w(\theta) = e^{\theta/\tau}, \qquad \theta \in (-\infty, 0], \tag{10}$$

is often used to model delay effects with a simple approximation. The effect of the past is fading away exponentially, and $\tau > 0$ is assumed as a measure of the delay in the system. The characteristic function

$$D(\lambda) = \lambda + c \int_{-\infty}^{0} e^{\theta/\tau}e^{\theta\lambda}d\theta = \lambda + c\frac{1}{\lambda + 1/\tau}$$

has only two zeros. They also come from the polynomial

$$\lambda^2 + \frac{\lambda}{\tau} + c. \tag{11}$$

The Routh-Hurwitz criterion implies that the trivial solution of (9) is asymptotically stable if and only if $c > 0$. The analysis of this DDE is easy, since with this exponential weight function the delay increases the dimension of the system by one only. It can be shown that the DDE (9) with the weight function (10) is equivalent to a second-order ODE. We differentiate (9) with respect to the time t and use partial integration to calculate

$$\ddot{x}(t) + c \int_{-\infty}^{0} e^{\theta/\tau} \dot{x}(t + \theta) d\theta = \ddot{x}(t) + cx(t) - \frac{c}{\tau} \int_{-\infty}^{0} e^{\theta/\tau} x(t + \theta) d\theta$$

$$= \ddot{x}(t) + \frac{1}{\tau} \dot{x}(t) + cx(t). \tag{12}$$

Its characteristic polynomial is just (11).

In oscillatory mechanical systems without viscous damping, the D-curves are usually lines in the parameter plane. The examples with finite and continuous delay such as

$$\ddot{x}(t) + c_0 x(t) - c_1 \int_{-1}^{0} w(\theta) x(t + \theta) d\theta = 0, \tag{13}$$

with the simplest weight function

$$w(\theta) \equiv 1, \qquad \theta \in [-1, 0], \tag{14}$$

have the characteristic function

$$D(\lambda) = \lambda^2 + c_0 - c_1 \int_{-1}^{0} e^{\lambda\theta} d\theta = \lambda^2 + c_0 - c_1 \frac{1 - e^{-\lambda}}{\lambda}, \qquad \lambda \neq 0, \tag{15}$$

where we define

$$D(0) = \lim_{\lambda \to 0} D(\lambda) = c_0 - c_1.$$

Since the D-curves (8) have the form

$$R(\omega) = -\omega^2 + c_0 - c_1 \frac{\sin \omega}{\omega} = 0, \qquad S(\omega) = c_1 \frac{1 - \cos \omega}{\omega} = 0,$$

the zeros of S can be given as $\omega = 2k\pi$, $k = 0, 1, \ldots$ or $c_1 = 0$, and the D-curves in the plane of c_0, c_1 are lines. Indeed

$$c_1 = 0 \quad \text{and} \quad c_0 > 0, \quad c_1 = c_0, \quad c_0 = 4k^2\pi^2.$$

The criteria (6) and (7), with $n = 1$ DOF, can be used to select the stability regions bordered by these lines. The first condition (6) clearly gives

$$c_0 \neq 4k^2\pi^2, \qquad k = 1, 2, \ldots. \tag{16}$$

If

$$0 < c_1 < c_0, \tag{17}$$

then

$$R(0) = c_0 - c_1 > 0 \quad \text{and} \quad \lim_{\omega \to \infty} R(\omega) = -\infty, \tag{18}$$

and the number r of the real positive zeros of R is odd, while S is positive at all the zeros ρ_k of R due to (16). Therefore

$$\sum_{k=1}^{r} (-1)^k \operatorname{sgn} S(\rho_k) = \sum_{k=1}^{r} (-1)^k = -1,$$

and the stability condition (7) is satisfied. If either $c_1 < 0$ or $c_1 > c_0$, then this condition is not satisfied. The stability chart of Figure 6.2 shows the shaded stability domains determined by the necessary and sufficient conditions (16)

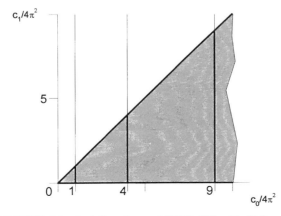

FIGURE 6.2. Stability chart of DDE (13) with (14).

and (17) for asymptotic stability. This example calls attention to the difficulty of selecting stability domains just by drawing the D-curves. Although the criteria (6) and (7) will not be checked and analyzed in the subsequent examples, we emphasize their importance in the construction of the stability charts.

If the DDE (13) is considered with the weight function

$$w(\theta) = -\frac{\pi}{2}\sin(\pi\theta), \qquad \theta \in [-1, 0], \tag{19}$$

then the characteristic function reads as

$$D(\lambda) = \lambda^2 + c_0 - c_1 \frac{\pi^2}{2} \frac{1 + e^{-\lambda}}{\lambda^2 + \pi^2}, \qquad \lambda \neq \pm i\pi,$$

with a continuous extension at $\pm i\pi$, namely with

$$D(\pm i\pi) = \lim_{\lambda \to \pm i\pi} D(\lambda) = -\pi^2 + c_0 \pm i\frac{\pi}{4}c_1.$$

This results in D-curves

$$R(\omega) = -\omega^2 + c_0 - c_1 \frac{\pi^2}{2} \frac{1 + \cos\omega}{\pi^2 - \omega^2} = 0, \quad S(\omega) = c_1 \frac{\pi^2}{2} \frac{\sin\omega}{\pi^2 - \omega^2} = 0,$$

where $S = 0$ can be solved in closed form, again giving $c_1 = 0$ or $\omega = j\pi$, $j = 0, 2, 3, \ldots$ (note $j \neq 1$). Substituting these into $R = 0$, we obtain the following lines as D-curves:

$$c_1 = 0 \text{ and } c_0 > 0, \quad c_1 = c_0, \quad \frac{1 + (-1)^j}{2}c_1 = -(j^2 - 1)(c_0 - j^2\pi^2), \qquad j = 2, 3, \ldots.$$

In the corresponding stability chart of Figure 6.3 the shaded stability regions selected by (6) and (7) are bordered by these lines. The ω values above the stability chart represent the critical angular frequencies at the corresponding stability limits.

A comparison of the stability charts in Figure 6.2 and 6.3 for the same DDE shows the strong influence of the weight function's shape on stability, although both weight functions satisfy

$$\int_{-1}^{0} w(\theta)\,d\theta = 1, \qquad \int_{-1}^{0} \theta w(\theta)\,d\theta = -\tfrac{1}{2}.$$

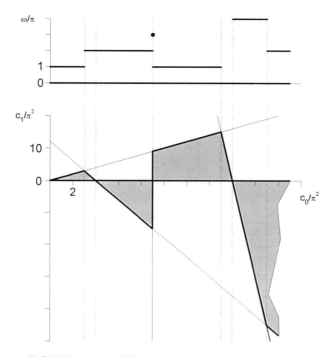

FIGURE 6.3. Stability chart of DDE (13) with (19).

The DDE

$$\ddot{x}(t) + b_0\dot{x}(t) + c_0x(t) - c_1x(t-1) = 0 \qquad (20)$$

can also be assumed as a special case of the DDE (13) when $b_0 = 0$, namely when there is no damping in the system, and the weight function is the so-called Dirac function at -1, namely $w(\theta) = \delta(1 + \theta)$. The D-curves calculated from the characteristic function

$$D(\lambda) = \lambda^2 + b_0\lambda + c_0 - c_1e^{-\lambda} \qquad (21)$$

can be given as

$$R(\omega) = -\omega^2 + c_0 - c_1\cos\omega = 0, \quad S(\omega) = b_0\omega + c_1\sin\omega = 0.$$

When $b_0 = 0$, these are equivalent to

$$c_1 = 0 \text{ and } c_0 > 0, \quad (-1)^jc_1 = c_0 - j^2\pi^2, \qquad j = 0, 1, \dots.$$

The equations above border the shaded stability regions in Figure 6.4,

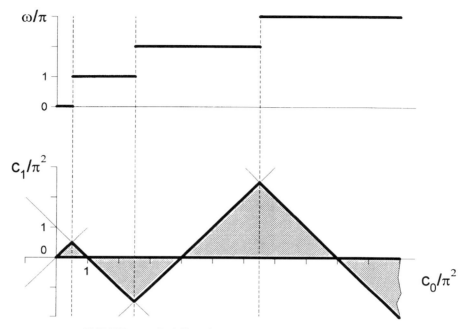

FIGURE 6.4. Stability charts of DDE (20) with $b_0 = 0$.

presenting a similar structure to that of Figure 6.3. Figure 6.5 also shows the stability regions when $b_0 = 1$. The damping increases the regions of stability, but the D-curves are not straight lines any more:

$$c_0 = \omega^2 - \frac{\omega}{\tan \omega}, \quad c_1 = -\frac{\omega}{\sin \omega}, \quad \omega \neq 0, \pi, 2\pi, \ldots.$$

Finally the stability chart of the DDE,

$$\frac{d^3}{dt^3} x(t) + b_0 \frac{d}{dt} x(t) - c_1 x(t-1) = 0, \tag{22}$$

is presented in Figure 6.6. Again the structure of this chart is the same, with the important difference that the stability domains are located only in the half-plane $c_1 < 0$. This odd order kind of DDE becomes important when continuous delay with the exponential weight function (10) is added to an oscillatory system already experiencing a discrete delay.

The stability chart also provides information about possible nonlinear vibrations in a nonlinear system whose linearization is the corresponding linear DDE. All but the stability limits $c_1 = c_0$ in the stability charts of Figures 6.3, 6.4, 6.5, and $c_1 = 0$ in Figure 6.6 refer to possible Hopf bifurcations with critical

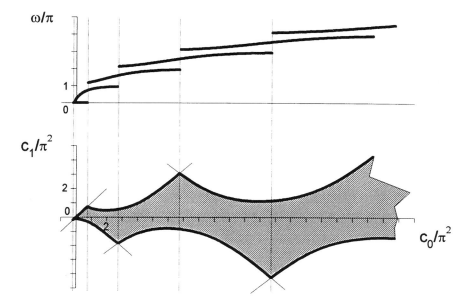

FIGURE 6.5. Stability charts of DDE (20) with $b_0 = 1$.

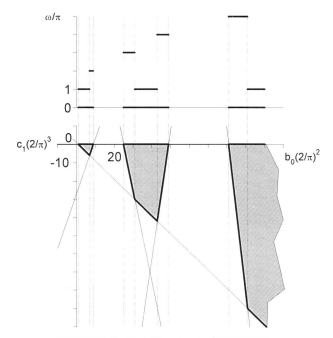

FIGURE 6.6. Stability chart of DDE (22).

characteristic roots $\pm i\omega$ presented above the charts. Stable or unstable periodic motions may appear there with angular frequency at about ω.

6.4   BASIC DDE MODELS OF REGENERATIVE MACHINE TOOL CHATTER

The basic idea of the regenerative effect has been explored in the literature (Tlusty and Spacek, 1954; Tobias, 1965), and experiments have clearly confirmed its existence. The machine tool has an elastic structure so that the tool and the workpiece can move relative to each other. Under stationary cutting conditions the chip thickness f, the chip width, and the cutting speed v have constant values prescribed by the designed technology. Any external or internal perturbation can cause a tool and a workpiece to start a damped vibration relative to each other, and as a result the surface of the workpiece becomes wavy. After one revolution of the workpiece (or the tool), the wavy surface will cause the chip thickness to vary at the tool's edge. Consequently the cutting force will vary and excite the stucture. This excitation frequency may seem to be very dangerous, since it is almost the same as the natural frequency of the structure. However, this phenomeon cannot be modeled as an excited vibration; it is rather a self-excited one where the cutting force is determined by the designed technological parameters, but its variation depends on the difference between the relative displacement of the tool and the workpiece at the given time instant and one round of the workpiece earlier. This special kind of self-excited vibration, where past effects also take place, is called *regenerative vibration*. In this sense the appearance of the regenerative vibration is a stability problem in a delayed oscillatory system, i.e., a damped oscillatory system with a delayed feedback or dead time.

Figure 6.7 shows the simplest, 1 DOF mechanical model of regenerative machine tool vibration in the planar case of orthogonal cutting. This model allows us to explain the basic stability problems and nonlinear vibrations arising in this system. In industrial applications the 1 DOF mechanical model is substituted by a thorough experimental modal analysis of the machine tool structure as is often presented in the specialist literature (e.g., Shi and Tobias, 1984). However, it is also true that the lowest natural frequencies are those most involved in regenerative vibrations and that low DOF mechanical models can still yield good quantitative results.

The zero value of the coordinate x of the tool edge position is set in a way that the x component F_x of the cutting force F is in balance with the spring's force, while the chip thickness f is just the prescribed value f_0. Then the equation of the tool motion is clearly

$$\ddot{x} + 2\kappa\alpha\dot{x} + \alpha^2 x = \frac{1}{m}\Delta F_x(f), \tag{23}$$

where $\alpha = \sqrt{s/m}$ is the *natural angular frequency* of the undamped free oscillating system and $\kappa = k/(2m\alpha)$ is the *relative damping factor*. The calcula-

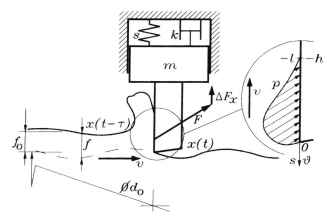

FIGURE 6.7. Mechanical model and cutting force variation.

tion of the cutting force variation ΔF_x requires an expression of the cutting force as a function of the technical parameters, primarily as a function of the chip thickness f which depends on the position x of the tool edge.

A simple empirical way to calculate the cutting force is by the Taylor approximation method. Applying the Taylor equations, we find that the cutting force F_x^T depends on the chip thickness, as shown in Figure 6.8, where the superscript T refers to the Taylor approximation. This is a degressive function usually given as a certain power of f that is less than 1. This simple method might be convenient in some technical design algorithms, but dynamical calculations require a power series formulation around f_0, such as

$$\Delta F_x^T(f) = F_x^T(f) - F_x^T(f_0) = \begin{cases} \Sigma_{j=1}^p k_j (\Delta f)^j & \text{if } \Delta f > -f_0, \\ -F_x^T(f_0) & \text{if } \Delta f \leqslant -f_0, \end{cases} \tag{24}$$

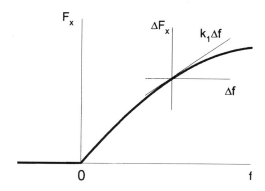

FIGURE 6.8. Cutting force and chip thickness relation.

where the chip thickness variation is

$$\Delta f = f - f_0. \tag{25}$$

The *cutting force coefficient*

$$k_1 = \frac{dF_x^T(f_0)}{df} \tag{26}$$

describes the linear approximation of the cutting force variation, and further coefficients of the power series come from

$$k_j = \frac{d^j F_x^T(f_0)}{j! df^j}, \qquad j = 2, 3, \ldots. \tag{27}$$

Note that the cutting force is zero for negative chip thickness ($f < 0 \Leftrightarrow \Delta f < - f_0$), which is of course something that physically does not exist. This is an important nonlinear part of the cutting force variation to be discussed later.

A better way to calculate the cutting force is to determine the stress resultant of the distributed force system P along the active face of the tool:

$$F_x(f) = \int_{-l}^{0} P_x(f, s) ds. \tag{28}$$

This is a challenging task of continuum mechanics. Although some early results can be found in the literature (Usui, Shirakashi, and Kitagawa, 1978), the most precise and detailed results has been published recently by Marusich and Ortiz (1996). Their finite-element algorithm accounts for inertia effects, contact and friction, heat generation and conduction, thermal softening, mechanical hardening, rate-sensitivity, brittle and ductile fracture, and fragmentation and permits chip morphologies to be a function of the technical parameters. This finite-element approach can trace the rigid body dynamics too, but the great amount of numerical work does not allow us to get an analytical picture of the regenerative effect nor to produce stability charts that summarize the effect of system and technical parameters on stability. As shown later, however, the variation of the distributed force system P on the active face of the tool has an important role in the dynamics of the system. The following approximation of the x component P_x of the force system combines the Taylor approximation of the cutting force F_x^T with an estimated shape function W (with unit 1/m):

$$P_x(f, s) = F_x^T(f) W(s), \qquad s \in [-l, 0],$$

where the origin of the local coordinate s is fixed to the tip of the tool and describes the distance (arc length) back along the active face having a length l, and

$$\int_{-l}^{0} W(s)ds = 1.$$

The estimation of W may be supported by the finite-element calculations mentioned above. Under stationary cutting conditions, $f \equiv f_0$ and

$$F_x(f_0) = \int_{-l}^{0} P_x(f_0, s)ds = F_x^T(f_0) \int_{-l}^{0} W(s)ds = F_x^T(f_0); \qquad (29)$$

in other words, the improved calculation of the cutting force results in the conventional Taylor description.

Let us introduce a long discrete time delay τ and a short continuous one h by

$$\tau = \frac{d_0 \pi}{v}, \quad h = \frac{l}{v}. \qquad (30)$$

A certain point on the surface of the workpiece needs time τ to meet the tool again, namely to travel one round with the cutting speed v along the circumference $d_0 \pi$ of the cylindrical workpiece of diameter d_0. The speed of the chip along the active face is proportional to the cutting speed v. For the sake of simplicity, let it also be v, which means that a certain particle of the chip needs the time h to slip along the active face.

The shape of the stress distribution can also be given in the time domain by introducing a "local time" $\theta = s/v$. This "local time" $\theta \in [-h, 0]$ is negative and shows how much earlier a certain particle of the chip was at the tip of the tool. The stress distribution function at this "local time" is denoted by w (with unit $1/s$):

$$w(\theta) = vW(v\theta) \Rightarrow \int_{-h}^{0} w(\theta)d\theta = 1. \qquad (31)$$

We then express the cutting force distribution in the x direction in the time domain by the global time t and local time θ:

$$p_x(t, \theta) = P_x(f(t, \theta), v\theta) = F_x^T(f(t, \theta)) \frac{1}{v} w(\theta), \qquad t \in [t_0, \infty), \theta \in [-h, 0]. \quad (32)$$

Above the active face of the tool, the chip thickness is approximated in the time domain by

$$f(t, \theta) = f_0 + x(t - \tau + \theta) - x(t + \theta), \qquad t \in [t_0, \infty), \, \theta \in [-h, 0]; \quad (33)$$

that is, the chip thickness at the tool tip is assumed to be

$$f(t, 0) = f_0 + x(t - \tau) - x(t).$$

As a result of the above relations (28), (29), (32), and (31), the power series (24), and also the chip thickness variation coming from (33), the cutting force variation in x direction reads as

$$\Delta F_x(f(t, .)) = F_x(f(t, .)) - F_x(f_0)$$

$$= \int_{-l}^{0} P_x\left(f\left(t, \frac{s}{v}\right), s\right) ds - F_x^T(f_0)$$

$$= \int_{-h}^{0} p_x(f(t, \theta), \theta) v d\theta - F_x^T(f_0)$$

$$= \int_{-h}^{0} (F_x^T(f(t, \theta)) - F_x^T(f_0)) w(\theta) d\theta$$

$$= \int_{-h}^{0} \left\{ \begin{array}{ll} \Sigma_{j=1}^{p} \dfrac{d^j F_x^T(f_0)}{j! df^j} (\Delta f(t, \theta))^j & \text{if } \Delta f(t, \theta) > -f_0 \\ -F_x^T(f_0) & \text{if } \Delta f(t, \theta) \leqslant -f_0 \end{array} \right\} w(\theta) d\theta$$

$$= \int_{-h}^{0} \left\{ \begin{array}{ll} \Sigma_{j=1}^{p} k_j (x(t-\tau+\theta) - x(t+\theta))^j & \text{if } x(t+\theta) < f_0 + x(t-\tau+\theta) \\ -F_x^T(f_0) & \text{if } x(t+\theta) \geqslant f_0 + x(t-\tau+\theta) \end{array} \right\} w(\theta) d\theta$$

$$(34)$$

$$\approx \int_{-h}^{0} k_1(x(t - \tau + \theta) - x(t + \theta)) w(\theta) d\theta. \qquad (35)$$

Substitution of the linearized cutting force variation (35) in the equation of motion (23) results in a linear DDE such as (2) in the scalar case with weights similar to (3) with respect to the past. Its stability analysis is presented in the next section. We obtain a nonlinear DDE suitable for Hopf bifurcation calculations if the Taylor series (24) truncated at the third degree ($p = 3$) is substituted in the calculation of the cutting force variation at (34). However, the global nonlinear behavior is also strongly determined by that part of the nonlinearity where the cutting force variation is constant, when the tool edge leaves the workpiece and the regenerative effect, namely the delay disappears from the system for a certain time period. The last section will deal with these nonlinear effects.

6.5 STABILITY OF CUTTING UNDER REGENERATIVE CONDITIONS

Even the linear model of regenerative machine tool vibrations exists in several more or less modified versions in the specialist literature. These modifications and improvements are needed to push the theoretical results closer to the experimental observations. Some of these results will be mentioned during the stability analysis of the model derived in the previous section.

Substitute the linear cutting force variation (35) into the differential equation (23) of the 1 DOF model:

$$\ddot{x}(t) + 2\kappa\alpha\dot{x}(t) + \alpha^2 x(t) + \frac{k_1}{m}\int_{-h}^{0} w(\theta)x(t+\theta)d\theta - \frac{k_1}{m}\int_{-\tau-h}^{-\tau} w(\tau+\theta)x(t+\theta)d\theta = 0.$$

(36)

The ratio of the short time delay h and the long delay τ is constant:

$$q = \frac{h}{\tau} = \frac{l}{d_0 \pi},$$

by their definitions in (30). From this point on, the long time delay τ will be kept as a parameter; namely

$$h = q\tau$$

is substituted, where τ is inversely proportional to the cutting speed v or to the angular velocity Ω of the workpiece:

$$\Omega = \frac{2\pi}{\tau}.$$

(37)

Its unit rad/s is converted in the stability charts below to rpm.

The trivial solution of (36) refers to the stationary cutting. When its stability is investigated, the D-curves coming from the characteristic function

$$D(\lambda) = \lambda^2 + 2\kappa\alpha\lambda + \alpha^2 + \frac{k_1}{m}D_0(\tau\lambda),$$

(38)

$$D_0(\tau\lambda) = \int_{-q\tau}^{0} w(\theta)e^{\lambda\theta}d\theta - \int_{-(1+q)\tau}^{-\tau} w(\tau+\theta)e^{\lambda\theta}d\theta,$$

(39)

are calculated as defined in (5):

$$R(\omega) = -\omega^2 + \alpha^2 + \frac{k_1}{m} R_0(\tau\omega) = 0, \tag{40}$$

$$R_0(\tau\omega) = \int_{-q\tau}^{0} w(\theta) \cos(\omega\theta)d\theta - \int_{-(1+q)\tau}^{-\tau} w(\tau + \theta) \cos(\omega\theta)d\theta, \tag{41}$$

$$S(\omega) = 2\kappa\alpha\omega + \frac{k_1}{m} S_0(\tau\omega) = 0, \tag{42}$$

$$S_0(\tau\omega) = \int_{-q\tau}^{0} w(\theta) \sin(\omega\theta)d\theta - \int_{-(1+q)\tau}^{-\tau} w(\tau + \theta) \sin(\omega\theta)d\theta. \tag{43}$$

In these equations of the D-curves, all the transcendental expressions are separated in the formulas of R_0 and S_0. They depend only on the product of the time delay τ and the critical frequency ω (which also serves as a parameter for the D-curves).

The stability charts are traditionally constructed in the plane of the cutting force coefficient k_1 and the angular velocity Ω of the workpiece (see (37)), since these parameters are proportional to the width of cut and the cutting speed, respectively, so the stability chart helps technology design in a somewhat direct way. The angular frequency ω of the vibration occurring at the loss of stability is also presented above the stability chart against the running speed Ω. The D-curves (40), (42) can be rearranged with respect to these technological parameters, and the stability limits can be plotted in the (Ω, k_1) plane with the following explicit expressions where the new parameter $\psi(=\tau\omega) \in \mathbb{R}^+$ is introduced:

$$\omega(\psi) = \alpha \left(-\kappa \frac{R_0(\psi)}{S_0(\psi)} + \sqrt{1 + \kappa^2 \frac{R_0^2(\psi)}{S_0^2(\psi)}} \right), \tag{44}$$

$$\Omega(\psi) = \frac{2\pi\omega(\psi)}{\psi}, \tag{45}$$

$$k_1(\psi) = -2\kappa\alpha m \frac{\omega(\psi)}{S_0(\psi)}, \tag{46}$$

where R_0 and S_0 are defined as in (41) and (43).

Four basic cases will be considered and discussed here. The first case is when the contact length l of the chip and tool is negligible relative to the circumference $d_0\pi$ of the workpiece. This can be modeled by choosing the Dirac function as the weight function:

$$w(\theta) = \delta(\theta). \tag{47}$$

Then the equation (23) of motion will contain only the long discrete delay τ:

$$\ddot{x}(t) + 2\kappa\alpha\dot{x}(t) + \alpha^2 x(t) + \frac{k_1}{m}(x(t) - x(t - \tau)) = 0,$$

and the expressions (41) and (43) simply give

$$R_0(\tau\omega) = 1 - \cos(\tau\omega),$$

$$S_0(\tau\omega) = \sin(\tau\omega).$$

Since

$$\frac{R_0(\psi)}{S_0(\psi)} = \frac{1 - \cos\psi}{\sin\psi} = \tan\frac{\psi}{2},$$

the parameter ψ can be eliminated from (44), and the stability limits (45) and (46) can be expressed in a more explicit form as a function of ω as follows:

$$\Omega(\omega) = \frac{\pi\omega}{j\pi - \mathrm{atn}\dfrac{\omega^2 - \alpha^2}{2\kappa\alpha\omega}}, \qquad j = 1, 2, \ldots, \tag{48}$$

$$k_1(\omega) = \frac{m}{2}\frac{(\omega^2 - \alpha^2)^2 + 4\kappa^2\alpha^2\omega^2}{\omega^2 - \alpha^2}. \tag{49}$$

Since $k_1 > 0$, the stability limit (49) already shows that the vibrations arising at the loss of stability will have vibration frequencies somewhat greater than the natural frequency of the system, that is $\omega > \alpha$. The stability chart in Figure 6.9 is constructed by means of the above D-curves in the same way as shown in the basic example (20) in Figure 6.5. The fixed parameters are $m = 50\,\mathrm{kg}$, $\kappa = 0.05$, $\alpha = 775\,\mathrm{rad/s}$.

It is important to observe that there exists a constant lower boundary of the stability limits that can easily be calculated from (49) as its minimum, where

$$\frac{dk_1}{d\omega}(\omega^*) = 0 \Rightarrow \omega^* = \alpha\sqrt{1 + 2\kappa}, \quad k_{1,\min} = k_1(\omega^*) = 2m\alpha^2\kappa(1 + \kappa).$$

This basic stability chart is well-known from the early books on machine tool vibration (Tobias, 1965). However, this stability chart has only been verified experimentally in the middle range of the cutting speed. The real cutting process shows somewhat better stability properties at low and high cutting speeds. To explain this experimental observation, Tobias introduced a so-called dynamic cuting theory where he inserted an additional damping in the equation of motion that was inversely proportional to the cutting speed.

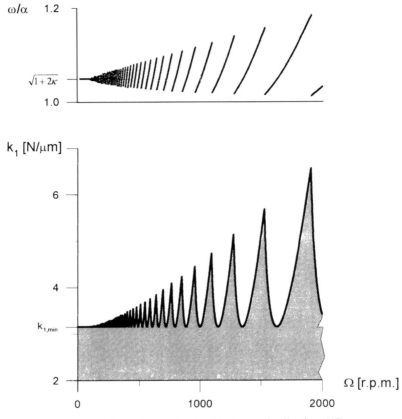

FIGURE 6.9. Stability chart with stress distribution (47).

This effect is very slight for turning, somewhat greater in the case of milling, and the strongest for drilling. Quantitative identification of such additional damping is difficult, since there is no universal method for correctly predicting regenerative vibrations at low and high speed. While introducing the idea of a complex cutting force coefficient may help get better quantitative agreement between theory and experiments (Tlusty, 1978), experimental results show that this cutting force coefficient cannot easily be identified as a function of cutting speed, frequency, or rake angle, for example, and the results are still not reliable for any kind of cutting.

For the weight functions W considered in this section, we furher call attention to the damping effect caused by the distributed force system P acting on the tool's active face. The damping effect becomes stronger as the tool-chip contact length l becomes greater relative to the distance $d_0\pi$ between two cutting edges, namely as q increases. Damping is minimized for turning and maximized for drilling.

Consider the case when the shape of the distributed cutting force system is approximated by an exponential function

$$W(s) = \frac{1}{l_0} \exp \frac{s}{l_0} \Rightarrow w(\theta) = \frac{v}{l_0} \exp\left(\frac{v}{l_0}\theta\right) = \frac{1}{q_0\tau} \exp \frac{\theta}{q_0\tau}, \qquad \theta \in (-\infty, 0], \quad (50)$$

where the contact length is infinite. Still, the length of this short-delay effect can be characterized by l_0, or by the ratio

$$q_0 = \frac{l_0/v}{d_0\pi/v} = \frac{h_0}{\tau}.$$

As in the basic example (9) and the equivalent equation (12), the equation of motion (36) can be transformed into a higher- (here third-) order system without a continuous time delay but still having the discrete delay τ:

$$q_0\tau \frac{d^3x}{dt^3}(t) + (1 + 2\kappa\alpha q_0\tau) \frac{d^2x}{dt^2}(t) + (2\kappa\alpha + \alpha^2 q_0\tau) \frac{dx}{dt}(t)$$

$$+ \left(\alpha^2 + \frac{k_1}{m}\right) x(t) - \frac{k_1}{m} x(t - \tau) = 0.$$

The stability of this equation can be analyzed in the same way as shown in Figure 6.6 for (22), though the stability limits will not be straight lines. The D-curves are constructed from (45) and (46) with

$$R_0(\psi) = \frac{1 - \cos\psi + q_0\psi \sin\psi}{1 + q_0^2\psi^2}, \quad S_0(\psi) = \frac{\sin\psi - q_0\psi(1 - \cos\psi)}{1 + q_0^2\psi^2}$$

using partial integration in Equations (41) and (43). The results are summarized in the stability chart of Figure 6.10 where the damping effect at low cutting speed clearly agrees with the experimental observations.

The stability chart gives an even more realistic result in the third case, where we consider

$$W(s) = \frac{1}{l}\left(1 + \cos\left(\frac{\pi}{l}s\right)\right), \ s \in [-l, 0] \Rightarrow w(\theta) = \frac{1}{q\tau}\left(1 + \cos\left(\frac{\pi}{q\tau}\theta\right)\right), \theta \in [-q\tau, 0].$$

$$(51)$$

Substituting this weight function in (41) and (43), we calculate the functions

$$R_0(\psi) = (1 - \cos\psi)\frac{\pi^2 \sin(q\psi)}{q\psi(\pi^2 - q^2\psi^2)} + \sin\psi \frac{\pi^2(1 - \cos(q\psi)) - 2q^2\psi^2}{q\psi(\pi^2 - q^2\psi^2)},$$

$$S_0(\psi) = \sin\psi \frac{\pi^2 \sin(q\psi)}{q\psi(\pi^2 - q^2\psi^2)} - (1 - \cos\psi)\frac{\pi^2(1 - \cos(q\psi)) - 2q^2\psi^2}{q\psi(\pi^2 - q^2\psi^2)},$$

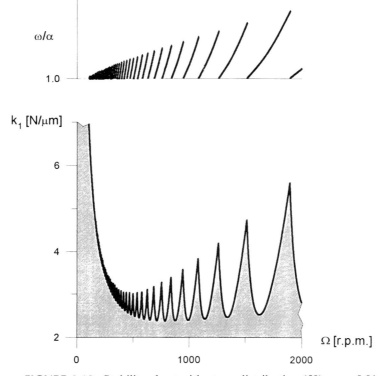

FIGURE 6.10. Stability chart with stress distribution (50), $q_0 = 0.01$.

and insert them into the stability chart formulas (45) and (46), as shown in Figure 6.11. In the figure improved stability properties can be seen at both low and high cutting speeds.

Finally we consider the stability chart for the weight function

$$W(s) = -\frac{\pi}{2l}\sin\left(\frac{\pi}{l}s\right), \ s \in [-l, 0] \Rightarrow w(\theta) = -\frac{\pi}{2q\tau}\sin\left(\frac{\pi}{q\tau}\theta\right), \ \theta \in [-q\tau, 0].$$

$$(52)$$

The corresponding stability chart is calculated using the functions

$$R_0(\psi) = \frac{\pi^2}{2}\left((1 - \cos\psi)\frac{1 + \cos(q\psi)}{\pi^2 - q^2\psi^2} + \sin\psi\frac{\sin(q\psi)}{\pi^2 - q^2\psi^2}\right),$$

$$S_0(\psi) = \frac{\pi^2}{2}\left(\sin\psi\frac{1 + \cos(q\psi)}{\pi^2 - q^2\psi^2} - (1 - \cos\psi)\frac{\sin(q\psi)}{\pi^2 - q^2\psi^2}\right),$$

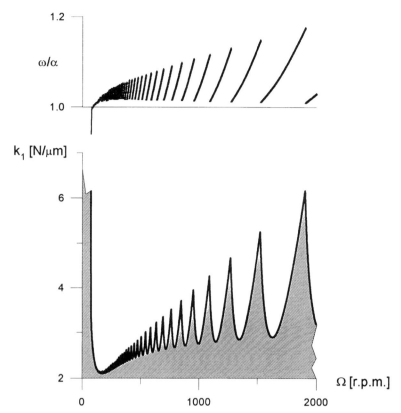

FIGURE 6.11. Stability chart with stress distribution (51), $q = 0.01$.

in the stability limits (45) and (46). Figure 6.12 presents a chart for low damping ($\kappa = 0.01$); it has a more complicated structure that includes a rich frequency content which may be very surprising from technical viewpoint. However, this chart may be useful in a realistic situation like drilling, when the ratio of the short delay to the long delay gets close to 1 and that maximum distributed cutting force is behind the tool edge because of the negative rake angle at the drill's chisel edge.

6.6 NONLINEAR REGENERATIVE VIBRATIONS OF MACHINE TOOLS

The stability charts for cutting under regenerative conditions often refer to complex linear stability properties that depend on technological and mechanical parameters. However, the practical applicability of these results is still limited. Apart of the uncertainty in identifying some of the parameters, this is

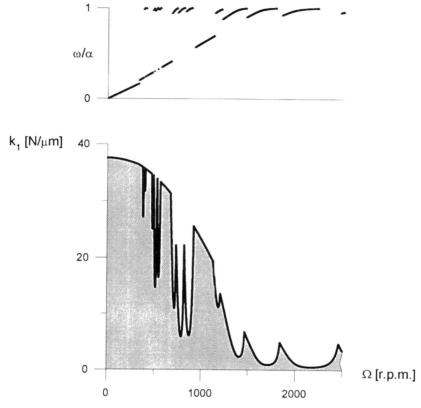

FIGURE 6.12. Stability chart with stress distribution (52), $q = 0.2$.

due to the fact that in the infinite-dimensional phase space the domain of attractivity of the stable stationary cutting remains very small. In other words, relatively small perturbations can push the system away from the otherwise asymptotically stable equilibrium, causing nonlinear vibrations. This has been proved experimentally (Shi and Tobias, 1984).

Two important reasons for nonlinear regenerative vibrations are discussed in this section. Both are related to the nonlinearity of the cutting force shown in Figure 6.8 and also in the cutting force variation formula (34). First, consider the "local" nonlinearity of a cutting force whose vibration amplitudes remain small, in the sense that the tool does not leave the workpiece:

$$x(t) < f_0 + x(t - \tau).$$

The stability charts of Figures 6.9, 6.10, 6.11, and 6.12 show that Hopf bifurcation occurs at the stability limits; that is, stable or unstable periodic motions exist around the equilibrium depending on the nature of the bifurcation, whether it is supercritical or subcritical. The approximate frequency of these vibrations is ω which is usually somewhat above the natural frequency of the system.

The Hopf bifurcation algorithm has been worked out in the literature (Hassard, Kazarinoff, and Wan, 1981), and the computer algebra helps in carrying out the tedious algebraic calculations. A third-degree approximation of the cutting force at the desired chip thickness f_0 will also contain second-degree terms, since the variation in the cutting force is not symmetric there. This makes the calculation difficult. Since the center manifold, the critical invariant two-dimensional surface embedded in the infinite-dimensional phase space, cannot be approximated by its tangent plane at the origin, determination of its second-degree approximation is also required. A closed-form algebraic approximation of the periodic motion is obtained, and stability is also determined. Note that the co-dimension two bifurcations that may occur refer to stable and unstable quasi-periodic oscillations (tori) in the phase space at frequencies where a discontinuity appears in ω above the stability charts. Such detailed calculations are available in the literature (Stépán and Haller, 1995; Campbell, Bélair, Ohira and Milton, 1993) for robotics and population dynamics.

We know from some numerical examples that the Hopf bifurcation in regenerative machine tool vibrations is subcritical, but there is no evidence or mathematical proof that it cannot be supercritical in other case. Some experiments (Shi and Tobias, 1984) refer to the existence of unstable periodic motions. Figure 6.13 presents a simplified bifurcation diagram for the case where the cutting force coefficient (or the width of cut) is the bifurcation parameter. The dashed curve refers to the unstable periodic motion or limit cycle, and this Hopf bifurcation calculation does not show any attractor around the unstable equilibrium (dashed line) or outside the unstable limit cycle.

There must be an attractor somewhere "outside". Consequently, since the

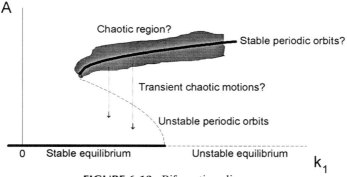

FIGURE 6.13. Bifurcation diagram.

Hopf bifurcation cannot describe that attractor, another, say, global, non-linearity must be considered. We will use the cutting force where the tool leaves the workpiece to serve this purpose:

$$x(t) \geqslant f_0 + x(t - \tau).$$

This is presented with respect to the cutting force variation in formula (34). This sort of nonlinearity calls for simulation to be used to get any information about the DDE model. The situation of an attractor arising from a global nonlinearity can easily be explained. If the system parameters are taken from the unstable region of the stability chart, or if they are from the stable region but the perturbations are great enough to push the state variables outside the unstable periodic motion, then the coordinate x increases, and the tool sooner or later leaves the workpiece. At this point the cutting force becomes zero (i.e., the cutting force variation in (34) becomes the constant $-F_x^T(f_0)$), and the regenerative effect is "switched off." This nonlinearity is so strong that the delay effect in the infinite-dimensional phase space is only valid for a certain region of this phase space. The trajectories spiralling outward will hit a surface from which they will jump onto the two-dimensional phase plane of the simple oscillating motion of the free tool outside the workpiece material. That motion is damped, of course, and the tool will soon return to the workpiece.

There are several ways in which the tool can react. The tool may arrive back from the phase plane into the infinite-dimensional phase space outside the unstable limit cycle. Then, as it goes out again, a series of switches may occur between the two dynamics: the infinite-dimensional delayed dynamics, and the 2-dimensional dynamics of the simple oscillator. As a result of these switches there may be some stable periodic or quasi-periodic motion, or even chaotic motion. Observation of the fractal-like surfaces of workpieces (Moon, 1994) has confirmed the possible existence of chaos in these systems. A third option for the system is that transient chaotic motion occurs, since after some chaotic jumps between the two dynamics, the trajectory may arrive back to the delayed dynamics inside the unstable limit cycle, and the temporary chaotic motion of the system will settle at stable stationary cutting. However, the length of this transient chaotic motion can vary stochastically and can be as long as the entire cutting process. The structure of the transient chaotic motion here is the same as that of the shimmy problem of towed wheels (Stépán, 1991), though in a higher-dimensional phase space.

6.7   CONCLUDING REMARKS

Regenerative vibrations are one of the most important problems of machine tool vibrations. The delay-differential equation models of regenerative vibrations of this chapter describe a particularly rich dynamics whose corresponding phase space is infinite dimensional. An interesting and complex view of these systems is further provided by a linear stability analysis of these models and by stability charts that in the space of the technical parameters set up fractal-like stability domains. In the case of nonlinear vibrations which refer to the existence of unstable periodic motions, there exist quasi-periodic motions, and transient chaotic motions may also occur in some parameter domains. These DDE models allow us to conclude that the prediction of machine tool vibration is difficult even if we have a reliable mechanical and mathematical model at hand.

ACKNOWLEDGMENTS

This research was partially supported by the Hungarian Scientific Research Foundation OTKA Grant No. 4-041, and the U.S.–Hungarian Science and Technology Program Grant No. 336.

REFERENCES

Campbell, S. A., Bélair, J., Ohira, T., and Milton, J. 1995. Complex dynamics and multistability in a damped oscillator with delayed negative feedback. *J. Dynamics Diff. Eqs.* **7**:213–236.

Hale, J. K. 1977. *Theory of Functional Differential Equations.* Springer-Verlag, New York.

Hassard, B. D., Kazarinoff, N. D., and Wan, Y. H. 1981. *Theory and Applications of Hopf Bifucations.* London Math. Soc. Lect. Note Series 41. Cambridge.

Kuang, Y. 1993. *Delay Differential Equations.* Academic Press, San Diego.

Marusich, T. D., and Ortiz, M. 1996. *J. Eng. Ind.*, forthcoming.

Minis, I., and Yanushevsky, R. 1993. Dynamics of milling. *J. Eng. Ind.* **115**:1–8.

Moon, F. C. 1994. Chaotic dynamics and fractals in material removal process. In Thompson, J. M. T., and Bishop, S. R., eds., *Nonlinearity and Chaos in Engineering Dynamics.* Wiley, New York, pp. 25–37.

Shi, H. M., and Tobias, S. A. 1984. A theory of finite amplitude machine tool instability. *Int. J. Mach. Tool Des. Res.* **24**:45–69.

Stépán, G. 1989. *Retarded Dynamical Systems.* Longman, Harlow, London.

Stépán, G. 1991. Chaotic motion of wheels. *Vehicle Systs. Dynamics* **20**:341–351.

Stépán, G., and Haller, G. 1995. Quasiperiodic oscillations in robot dynamics. *Nonlin. Dynamics* **8**:513–528.

Tlusty, J. 1978. Analysis of the state of the research in cutting dynamics. *Ann. CIRP* **27**:583–589.

Tlusty, J., and Spacek, L. 1954. *Self-Excited Vibrations on Machine Tools* (in Czech). Nakl CSAV, Prague.

Tobias, S. A. 1965. *Machine Tool Vibrations.* Blackie, London.

Usui, E., Shirkashi, T., and Kitagawag, T. 1978. Analytical prediction of three dimensional cutting process. *J. Eng. Ind.* **100**:236–243.

7

APPLICATIONS OF
PERTURBATION
METHODS TO TOOL
CHATTER DYNAMICS

ALI H. NAYFEH, CHAR-MING CHIN, and JON PRATT

7.1 INTRODUCTION

The role of perturbation methods and bifurcation theory in predicting the stability and complicated dynamics of machining are discussed in this chapter, using a nonlinear single-degree-of-freedom model. An introduction to these methods may be found in the text by Nayfeh and Balachandran (1995). According to linear theory, disturbances decay with time and hence chatter does not occur if the cutting width $w < w_c$ and disturbances grow exponentially with time, and hence chatter occurs if $w > w_c$. In other words, as w increases beyond w_c, a Hopf bifurcation occurs leading to the birth of a limit cycle. To ascertain the stability of this limit cycle, and hence whether the bifurcation is supercritical or subcritical, we use the method of multiple scales to determine the normal form of the equation near the Hopf bifurcation and include the effects of the quadratic and cubic nonlinearities. To compute the evolution of the limit cycles, we implement a method of harmonic balance using a symbolic manipulator, namely MAPLE. Then we generate bifurcation diagrams describing the variation in the amplitude of the fundamental harmonic with the width of cut. Using a combination of Floquet theory and Hill's determinant, again implemented using MAPLE, we ascertain the stability of the limit cycles. There

Dynamics and Chaos in Manufacturing Processes, Edited by Francis C. Moon.
ISBN 0-471-15293-5 © 1998 John Wiley & Sons, Inc.

are two cyclic-fold bifurcations, resulting in large-amplitude limit cycles, hysteresis, jumps, and subcritical instability. As the width of cut w increases, the periodic solutions undergo a secondary Hopf bifurcation, leading to a two-period quasi-periodic motion (two torus). The periodic and quasi-periodic solutions are verified by numerical simulation. As w increases further, each torus breaks down, resulting in chaotic motion.

Increasingly manufacturing engineers need accurate predictions and appraisals of machine-tool performance to select operating parameters, such as spindle speed and tool feed, that lead to the best possible part created in the most efficient manner. To this end we look at how perturbation methods may be employed to reveal the dynamics of self-excited oscillations that occur during single-point machining of metals. This phenomenon, commonly known as chatter, is a dynamic instability inherent in the cutting process.

Chatter limits the operating regimes available to the machine-tool user, since combinations of spindle speeds and tool feeds that make cutting unstable must be avoided to prevent damage to the machine tool, the workpiece, or the operator. Vibrations of any kind in the machine-tool system cause poor surface finish, introducing machining errors. Typical sources of vibration are the spindle drive, the work environment, moving and sliding members, and the cutting process itself (chatter). In particular, chatter may increase the dynamic loading of the machine-tool structure to the point where tools break or the workpiece is damaged beyond salvage.

Efforts to understand chatter have a long history as evidenced by the literature. Fundamental chatter theories were first proposed by Tobias and Tlusty and published in their excellent monographs (Tobias, 1965; Koeingsberger and Tlusty, 1970). These works have provided the framework within which the problem is usually formulated.

Three mechanisms of chatter—namely regenerative effect, mode coupling or flutter-type instability, and velocity-dependent effect—are often considered. The regeneration mechanism occurs whenever cuts overlap and the cut produced at time t leaves small "waves" in the material that are regenerated with each subsequent pass of the tool. Mode coupling results whenever the relative vibration between the tool and the workpiece exists simultaneously in two directions in the plane of the cut. In this case the tool traces out an elliptic path that varies the depth of cut in such a fashion as to feed the coupled modes of vibration. This mechanism is similar to the aeroelastic problem of flutter, where a structural instability arises due to the dependence of the dynamic stiffness on the aerodynamic load. Finally the velocity-dependent effect is the result of the variation of the cutting force with the cutting speed having a negative slope (Arnold, 1946). The velocity-dependent effect is a source of negative damping similar to that in the Rayleigh and van der Pol oscillators (Nayfeh and Mook, 1979).

Despite the general acceptance of the aforementioned chatter mechanisms, linear stability theories built around them have achieved limited success, and precise methods of predicting the stability of cutting remain elusive. New approaches for analyzing the problem emerge regularly in the literature (Minis et al., 1990). Increasingly authors are considering the nonlinear aspects of the problem (Grabec, 1986, 1988; Moon, 1994; Lin and Weng, 1990, 1991; Tlusty and Ismail, 1982). A review of this literature reveals that nonlinearity may arise in the structural stiffness, the cutting force, and the friction induced at the tool-chip interface.

Many of these studies were concerned primarily with demonstrating the potential for chaos. We suggest that a more practical concern is whether or not the nonlinearity results in subcritical or supercritical instability.

Among the first to recognize this fact were Hanna and Tobias (1974), who observed a jump phenomenon in a face milling operation. They reported that for the same cutting parameters both stable cutting and large-amplitude chatter coexisted. To investigate this behavior, they developed a nonlinear single-degree-of-freedom model. The model was analyzed using a two-term harmonic balance; it revealed that cuts thought to be stable according to linear theory were in fact prone to large-amplitude motions due to the nonlinearities.

Nayfeh, Chin, and Pratt (1997) used the modern methods of nonlinear dynamics (the method of multiple scales, harmonic balance, Floquet theory, and numerical methods) to investigate more thoroughly the impact of time delay and nonlinearity on the dynamics of the single-degree-of-freedom model of Hanna and Tobias (1974). They demonstrated that time-delay and nonlinear stiffness effects are in fact sufficient to cause a single-degree-of-freedom non-linear oscillator to exhibit limit-cycle, quasi-periodic, and chaotic dynamics. Pratt and Nayfeh (1996) explored the complicated dynamics of the model of Hanna and Tobias (1974) using analog computer simulations. The experimental results validate the theoretical results of Nayfeh, Chin, and Pratt (1997). Here we describe applications of perturbation methods to tool chatter dynamics by using the following model of Hanna and Tobias:

$$\ddot{x} + 2\xi\dot{x} + p^2(x + \beta_2 x^2 + \beta_3 x^3) = -p^2 w[x - x_\tau + \alpha_2(x - x_\tau)^2 + \alpha_3(x - x_\tau)^3],$$

$$\tag{1}$$

$$x_\tau = x(t - \tau), \tag{2}$$

where x is the displacement normal to the machined surface, damping is of the hysteretic type and ξ is inversely proportional to the chatter frequency ω, w is the effective width of cut, p is the natural frequency of the cutter, β_2 and β_3 are two constants describing the nonlinear stiffness of the machine tool, and α_2 and α_3 are nonlinear constant coefficients of the cutting force function. This model falls somewhere between that of Minis et al. (1989), who completely linearized the model of Wu and Liu (1985), and that of Lin and Weng (1990), who kept all the nonlinearities and used numerical integration to analyze the system. The model does not include part stiffness effects, though they could be added. Arguably the physics of the problem pushes one toward ever more sophisticated models with more degrees of freedom, models of tool run out, and finite sources of power, and so on. However, the model of Hanna and Tobias incorporates many of the important mechanisms that drive the local dynamics, namely the regenerative effect and the structural and cutting force nonlinearities.

In this review we present numerical results obtained for the model of Hanna and Tobias, using their experimentally determined parameter values (Hanna and Tobias, 1974): $p = 1088.56 \, \text{rad/s}$, $\xi = 24792/\omega \, \text{rad}^2/\text{s}$, $\beta_2 = 479.3 \, 1/\text{in}$, $\beta_3 = 264500 \, 1/\text{in}^2$, $\alpha_2 = 5.668 \, 1/\text{in}$, and $\alpha_3 = -3715.2 \, 1/\text{in}^2$.

We begin with the linear analysis to determine the local stability boundary. Then we use the method of multiple scales to determine the time evolution of the amplitudes and phases of the periodic solutions (limit cycles) created as a result of the instability. This analysis yields the so-called normal form, which is used to ascertain whether the created limit cycles are stable or unstable and hence whether the accompanying bifurcation is supercritical or subcritical. To study the evolution of periodic solutions and their stability for progressively larger cut widths, we implement the method of harmonic balance and Floquet theory by using MAPLE. Finally we compare these predictions with the results of numerical integration and then explore the route to chaos using digital simulation.

7.2   LINEAR PROBLEM

The linearized form of Equation (1) can be written as

$$\ddot{x} + 2\xi\dot{x} + p^2 x + p^2 w(x - x_\tau) = 0. \tag{3}$$

To study the neutral stability, we substitute

$$x = a \cos \omega t \tag{4}$$

into Equation (3), set each of the coefficients of $\cos \omega t$ and $\sin \omega t$ equal to zero, and obtain

$$p^2 - \omega^2 + p^2 w(1 - \cos \omega\tau) = 0, \tag{5}$$

$$2\xi\omega + p^2 w \sin \omega\tau = 0. \tag{6}$$

We establish the linear stability boundary in Figure 7.1 by solving Equations (5) and (6) for w, τ, and ω. It follows from Equation (6) that when $w < 2\xi\omega/p^2$, which is 0.0418 in the current study, the cutting process remains stable. As shown in Figure 7.1, the instability occurs at the critical cutting width ($w = w_c \geqslant 0.0418$), which depends on the time delay τ. For a given τ and ξ, x decays with time, and hence chatter does not occur if $w < w_c$, x grows exponentially with time and hence chatter occurs if $w > w_c$. Adding the

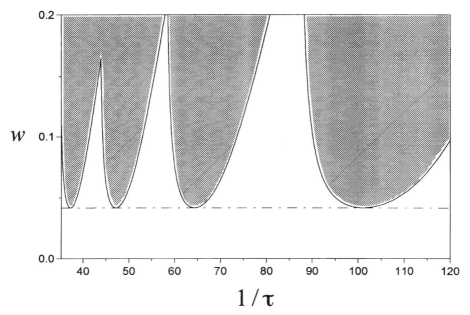

FIGURE 7.1. Linear stability diagram with w being the effective width of cut and τ being the delay time. Shaded area represents the region of unstable cutting process.

nonlinearity limits the amplitude to a finite value. Moreover, as shown in Section 7.3, adding the nonlinearity produces a so-called subcritical instability, which results in chatter for values of w far below w_c in the presence of large disturbances. Hence, in the absence of large disturbances, increasing w past the critical value w_c results in the system losing stability from a static cutting to a dynamic one. Such an instability is called a *Hopf bifurcation*. In the next section we investigate the normal form for this bifurcation to ascertain whether the limit cycles created as a result of the bifurcation are stable or unstable.

7.3 NORMAL FORM NEAR HOPF BIFURCATION

The nonlinearities in Equation (1) include the nonlinear stiffness of the machine tool as well as the nonlinear cutting force generated in the cutting process. According to the linear analysis, a Hopf bifurcation occurs, leading to the birth of a limit cycle as w increases past w_c. The qualitative behavior in the neighborhood of this bifurcation can be predicted by studying the corresponding normal form. Due to the presence of both cubic and quadratic nonlinearities, the perturbation analysis can be performed using either the method of multiple scales or the generalized method of averaging (Nayfeh, 1973, 1981). In this review we use a multiple-scales approach. To this end we introduce a fast time scale $T_0 = t$ and a slow time scale $T_2 = \varepsilon^2 t$ and seek a third-order expansion in the form

$$x = \varepsilon x_1(T_0, T_2) + \varepsilon^2 x_2(T_0, T_2) + \varepsilon^3 x_3(T_0, T_2) + \cdots. \qquad (7)$$

Moreover we perturb the control parameter (effective width of cut), and let

$$w = w_c + \varepsilon^2 w_2. \qquad (8)$$

Substituting Equations (7) and (8) into (1) and equating coefficients of like powers of ε, we obtain

$$D_0^2 x_1 + 2\xi D_0 x_1 + p^2 x_1 + p^2 w_c(x_1 - x_{1\tau}) = 0 \qquad \text{(order } \varepsilon), \qquad (9)$$

$$\begin{aligned} D_0^2 x_2 &+ 2\xi D_0 x_2 + p^2 x_2 + p^2 w_c(x_2 - x_{2\tau}) \\ &= -p^2 \beta_2 x_1^2 - p^2 w_c \alpha_2(x_1 - x_{1\tau})^2 \qquad \text{(order } \varepsilon^2), \end{aligned} \qquad (10)$$

$$\begin{aligned} D_0^2 x_3 &+ 2\xi D_0 x_3 + p^2 x_3 + p^2 w_c(x_3 - x_{3\tau}) \\ &= -2D_0 D_2 x_1 - 2\xi D_2 x_1 - p^2 w_2(x_1 - x_{1\tau}) \\ &\quad - 2p^2 \beta_2 x_1 x_2 - 2p^2 w_c \alpha_2(x_1 - x_{1\tau})(x_2 - x_{2\tau}) \\ &\quad - p^2 \beta_3 x_1^3 - p^2 w_c \alpha_3(x_1 - x_{1\tau})^3 \qquad \text{(order } \varepsilon^3). \end{aligned} \qquad (11)$$

The solution of Equation (9) can be expressed as

$$x_1 = A(T_2)e^{i\omega_c T_0} + cc \quad \text{and} \quad x_{1\tau} = A(T_2)e^{i\omega_c(T_0 - \tau)} + cc \qquad (12)$$

where ω_c is the chatter frequency at the Hopf bifurcation. Substituting Equation (12) into Equation (10) and solving for x_2, we obtain

$$x_2 = -p^2 A^2 \Gamma_1 e^{2i\omega_c T_0} - 2\Gamma_2 A\bar{A} - p^2 \bar{A}^2 \bar{\Gamma}_1 e^{-2i\omega_c T_0}, \qquad (13)$$

where

$$\Gamma_1 = \frac{\beta_2 + w_c \alpha_2(1 - e^{-i\omega_c \tau})^2}{p^2 - 4\omega_c^2 + 4i\omega_c \xi + p^2 w_c(1 - e^{-2i\omega_c \tau})}, \qquad (14)$$

$$\Gamma_2 = \beta_2 + w_c \alpha_2(1 - e^{-i\omega_c \tau})(1 - e^{i\omega_c \tau}). \qquad (15)$$

Substituting Equations (12) and (13) into Equation (11) and eliminating the terms that lead to secular terms, we obtain the solvability condition

$$2(\xi + i\omega_c)A' + p^2 w_2(1 - e^{-i\omega_c \tau})A + \Lambda A^2 \bar{A} = 0, \qquad (16)$$

where

$$\Lambda = \Lambda_r + i\Lambda_i$$
$$= -2p^2\beta_2(p^2\Gamma_1 + 2\Gamma_2) + 3p^2\beta_3 - 2p^4w_c\alpha_2\Gamma_1(1 - e^{i\omega_c\tau})(1 - e^{-2i\omega_c\tau})$$
$$+ 3p^2w_c\alpha_3(1 - e^{-i\omega_c\tau})^2(1 - e^{i\omega_c\tau}), \qquad (17)$$

and Λ_r and Λ_i are real constants. Introducing the polar transformation

$$A = \tfrac{1}{2}ae^{i\beta} \qquad (18)$$

into Equation (16) and separating real and imaginary parts, we obtain the normal form

$$a' = c_1 w_2 a + c_2 a^3, \qquad (19)$$

$$a\beta' = c_3 w_2 a + c_4 a^3, \qquad (20)$$

where

$$c_1 = -\frac{p^2}{2(\xi^2 + \omega_c^2)}\left[\xi(1 - \cos\omega_c\tau) + \omega_c\sin\omega_c\tau\right] \qquad (21)$$

$$c_2 = -\frac{1}{8(\xi^2 + \omega_c^2)}(\xi\Lambda_r + \omega_c\Lambda_i) \qquad (22)$$

$$c_3 = -\frac{p^2}{2(\xi^2 + \omega_c^2)}\left[\xi\sin\omega_c\tau - \omega_c(1 - \cos\omega_c\tau)\right] \qquad (23)$$

$$c_4 = -\frac{1}{8(\xi^2 + \omega_c^2)}(\xi\Lambda_i - \omega_c\Lambda_r). \qquad (24)$$

The qualitative behavior near a Hopf bifurcation can be determined by the sign of c_2. The bifurcation is supercritical when $c_2 < 0$ and subcritical when $c_2 > 0$. For example, a supercritical Hopf bifurcation occurs at $(\tau, w_c, \omega_c) = (1/75, 0.117072, 1205.51)$ because $(c_1, c_2, c_3, c_4) = (159, -787737, 953, -2608438)$. A supercritical Hopf bifurcation is also predicted at $(\tau, w_c, \omega_c) = (1/60, 0.075037, 1095.48)$ because $(c_1, c_2, c_3, c_4) = (300, -195328, 98, 3678391)$. Hence local disturbances decay for $w < w_c$ and result in small limit cycles (periodic motions) for $w > w_c$.

7.4 PERIODIC SOLUTIONS AND THEIR STABILITY

To construct analytically the periodic solutions of Equation (1), we note that they can be expressed in Fourier series. Hanna and Tobias (1974) truncated the series by using two terms. It turns out that two terms in the series may not be enough. Nayfeh, Chin, and Pratt (1997) found that including up to the third harmonics yields accurate results when compared with the results of numerical integration. Thus we let

$$x = a_0 + a_1 \cos \omega t + \sum_{n=2}^{N} (a_n \cos n\omega t + b_n \sin n\omega t), \qquad (25)$$

where N is a finite integer. Substituting Equation (25) into Equation (1) and equating the coefficient of each harmonic on both sides, we obtain a system of nonlinear algebraic equations. The algebra can be carried out by using a symbolic manipulator. We numerically calculate the roots of this system of equations and hence ω and the a_i and b_i. We choose N so that the results do not change on increasing it. Due to the lengthy expressions in these equations, we only show the numerical results in the figures.

To determine the stability of a periodic solution $x_0(t)$, we use Floquet theory (Nayfeh and Mook, 1979; Nayfeh and Balachandran, 1995). We first perturb the periodic solution $x = x_0(t)$ by introducing a disturbance term $u(t)$, and we obtain

$$x(t) = x_0(t) + u(t). \qquad (26)$$

Substituting Equation (26) into Equation (1) and linearizing in the disturbance u, we obtain

$$\ddot{u} + 2\xi\dot{u} + p^2(1 + 2\beta_2 x_0 + 3\beta_3 x_0^2)u$$
$$= -p^2 w[1 + 2\alpha_2(x_0 - x_{0\tau}) + 3\alpha_3(x_0 - x_{0\tau})^2](u - u_\tau). \qquad (27)$$

According to Floquet theory, Equation (27) admits solutions of the form

$$u(t) = e^{\gamma t}\phi(t), \qquad (28)$$

where $\phi(t)$ is a periodic function with period $T = 2\pi/\omega$, which is equal to the period of $x_0(t)$. Substituting Equation (28) into Equation (27) yields

$$\ddot{\phi} + 2\gamma\dot{\phi} + \gamma^2\phi + 2\xi\gamma\phi + 2\xi\dot{\phi} + p^2(1 + 2\beta_2 x_0 + 3\beta_3 x_0^2)\phi$$
$$= -p^2 w[1 + 2\alpha_2(x_0 - x_{0\tau}) + 3\alpha_3(x_0 - x_{0\tau})^2](\phi - e^{-\gamma\tau}\phi_\tau). \qquad (29)$$

To solve Equation (29), we expand ϕ in a Fourier series and keep up to the Nth harmonics for consistency with Equation (25). Thus we let

$$\phi = b_0 + \sum_{n=1}^{N} (c_n \cos n\omega t + d_n \sin n\omega t). \qquad (30)$$

Substituting Equation (30) into Equation (29) and equating the coefficient of each harmonic on both sides, we obtain a system of linear homogeneous algebraic equations governing the c_m, d_m, and γ. Setting the determinant of the coefficient matrix of this system equal to zero, we obtain the so-called Hill's determinant governing γ. This determinant is more complex than a common polynomial form due to the time delay τ. Again the algebra can be carried out by using a symbolic manipulator.

It follows from Equation (28) that a given periodic solution $x_0(t)$ is asymptotically stable if and only if all of the eigenvalues γ lie in the left half of the complex plane, and it is unstable if at least one eigenvalue lies in the right half-plane. Starting with a width of cut w for which periodic solutions are stable and then increasing w, we find that these periodic solutions will undergo a bifurcation and hence lose stability. The response subsequent to the bifurcation depends on the manner in which the eigenvalues move from the left to the right half-plane. If a real eigenvalue moves from the left to the right half-plane along the real axis, the system response will jump to another solution, which in this case may be periodic or trivial. On the other hand, when a pair of complex conjugate eigenvalues moves transversely from the left to the right half-plane, the resulting response will be either a two-period quasi-periodic motion (two torus) or a periodic motion with a large-period (phase-locked) motion.

7.5   NUMERICAL RESULTS

7.5.1   Convergence

As a first step we compare in Figure 7.2 variation of the analytically obtained amplitude a of the fundamental harmonic with the width of cut w by using two-term (i.e., $N = 1$) and six-term (i.e., $N = 3$) approximations in the Fourier series when $\tau = 1/75$. Such a variation is called a bifurcation diagram. Except for the small-amplitude limit cycles for values of w near w_c, as shown in the inset, the two approximations yield results that are very close. However, whereas the six-term approximation predicts a secondary Hopf bifurcation at $w \approx 0.141$, the two-term approximation predicts stable limit cycles. When $\tau = 1/60$, the results in Figure 7.3 show that the two-term approximation predicts qualitatively erroneous results; it predicts a subcritical rather than a supercritical Hopf bifurcation, as shown in the inset.

Increasing the number of terms in the harmonic balance beyond six did not produce any qualitative or quantitative change in the response. Therefore six terms were used to generate all analytical results presented in the remainder of this chapter.

Next we validate the analytical results by numerically integrating Equation (1). For the numerical integration, we use a fifth-order Runge-Kutta scheme. Then we start with some initial conditions and, as in the actual cutting process, let $x_\tau = 0$ for $0 \leqslant t < \tau$ and switch on x_τ for $t \geqslant \tau$. The integration is carried out long enough for the transients to die out. The numerically obtained

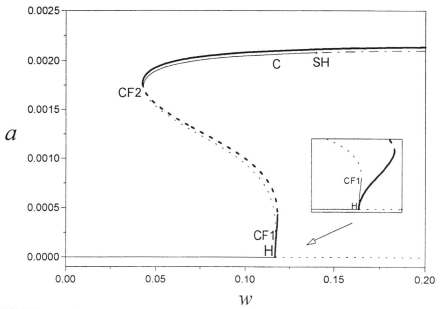

FIGURE 7.2. Comparison of the nonlinear response curves obtained by using the method of harmonic balance when $\tau = 1/75$: Thin lines obtained with a six-term approximation, and thick lines obtained with a two-term approximation. Both show supercritical Hopf bifurcations. Solid lines denote stable trivial or periodic solutions, dotted lines denote unstable trivial or periodic solutions, and dotted-dashed line denotes the region of quasi-periodic and chaotic solutions.

periodic solutions are very close to those obtained analytically so much so that they are indistinguishable when plotted together. Therefore we show in Figure 7.4 two-dimensional projections of the phase portraits of the periodic solutions (limit cycles) obtained with the two approaches for the two widths of cut $w = 0.06$ and 0.136 when $\tau = 1/75$. The agreement is excellent and indicates that the harmonic-balance solution is accurate for periodic motions. Numerical integration is used exclusively once the periodic solutions lose stability.

7.5.2   Subcritical Instability

It follows from the bifurcation diagram shown in Figure 7.2 that the trivial solution is stable for $w < w_c$ and that it is a sink; the trivial solution is unstable for $w > w_c$ and it is a saddle. The nontrivial solutions represented by solid curves are stable and correspond to stable limit cycles. A limit cycle is an isolated periodic solution. The nontrivial solutions represented by the dotted curves correspond to unstable limit cycles with a real eigenvalue γ lying in the right half-plane; hence they correspond to limit cycles of the saddle type. The

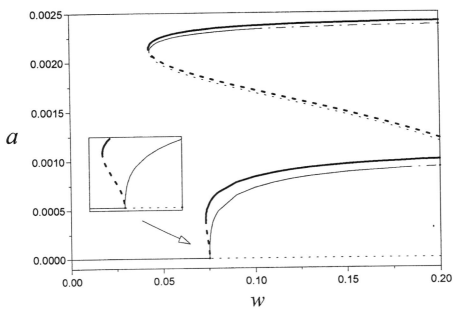

FIGURE 7.3. Comparison of the nonlinear response curves obtained by using the method of harmonic balance when $\tau = 1/60$: Thin lines obtained with a six-term approximation, and thick lines obtained with a two-term approximation. The two-term approximation erroneously predicts a subcritical rather than a supercritical Hopf bifurcation. Solid lines denote stable trivial or peridic solutions, dotted lines denote unstable trivial or periodic solutions, and dotted-dashed line denotes the region of quasi-periodic and chaotic solutions.

nontrivial solutions represented by the dotted-dashed curve correspond to unstable limit cycles with a pair of complex conjugate eigenvalues lying in the right half-plane; the response in this case is either a two-period quasi-periodic motion or a phase-locked motion or a chaotic motion. There are four bifurcations labeled as H, $CF1$, $CF2$, and SH. In the neighborhood of H, there exist a branch of stable trivial solutions for $w < w_c$, a branch of unstable trivial solutions, and a branch of stable limit cycles for $w > w_c$. Such a bifurcation is called *supercritical*. The lower branch of stable limit cycles meets the branch of saddle limit cycles at $CF1$, resulting in the destruction of both limit cycles. Such a bifurcation is called *cyclic-fold*. Similarly the upper branch of stable limit cycles meets the branch of saddle limit cycles at $CF2$, resulting in the destruction of both limit cycles in a cyclic-fold bifurcation. In the neighborhood of SH, there is a branch of stable limit cycles on one side and a branch of unstable limit cycles and a branch of stable two-period quasi-periodic motions on the other side. Such a bifurcation is called *supercritical secondary Hopf bifurcation*.

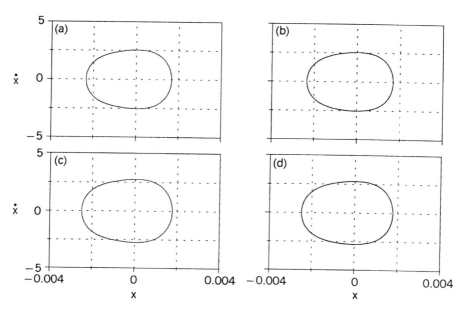

FIGURE 7.4. Comparison of the phase portraits obtained using the method of harmonic balance (a, c) with those obtained using numerical integration (b, d) when $\tau = 1/75$ for (a, b) $w = 0.06$ and (c, d) $w = 0.136$.

In the absence of large disturbances, performing an experiment starting with a small value of w and slowly increasing w, one finds that the response is trivial and hence that there is no chatter until $w = w_c$. Increasing w past w_c, one finds that the trivial solution loses stability, giving way to small limit cycles. As w is increased further, the limit cycle deforms and grows until it collides with an unstable limit cycle (the dotted branch at point $CF1$), resulting in the destruction of both limit cycles. Consequently the system response jumps to point C, which corresponds to a large-amplitude limit cycle. Increasing w past point C results in a slow deformation and growth of the limit cycle until the secondary Hopf bifurcation SH is reached. Starting to the left of SH and slowly decreasing w, one finds that the response continues to be a large-amplitude limit cycle past $CF1$ and H until the cyclic-fold bifurcation $CF2$ is reached. Decreasing w below $CF2$ results in the destruction of the limit cycle and a jump to the trivial solution. We note that in between $CF2$ and H, the system response may be trivial (no chatter) or a large-amplitude limit cycle (chatter with a large amplitude), depending on the initial conditions. This phenomenon is usually referred to as a *subcritical instability*. As a consequence the nonlinearity reduces the instability limit $w = w_c \approx 0.117072$ predicted by the linear theory, to $w \approx 0.0431$, corresponding to $CF2$, a reduction of about 63%. The subcritical instability was observed experimentally by Hanna and Tobias (1974).

7.5.3   Influence of Time Delay

In Figure 7.5 we show variation of the amplitude a of the fundamental harmonic with the width of cut w for the four time delays $\tau = 1/60$, $1/64.4$, $1/66$, and $1/75$. The first time delay corresponds to the left side, whereas the fourth time delay corresponds to the right side of the lobe in Figure 7.1. We note that all Hopf bifurcations are supercritical, leading to the birth of small-amplitude limit cycles. When $\tau = 1/60$, the created limit cycles grow slowly with increasing w until they lose stability via a secondary Hopf bifurcation, resulting in quasi-periodic solutions. As τ decreases to $1/64.4$, again the created limit cycles grow with increasing w but at a faster rate than in the case $\tau = 1/60$, and hence they suffer a secondary Hopf bifurcation for a smaller value of w. As τ decreases slightly, the created limit cycles grow at an even faster rate with w and lose stability via a cyclic-fold bifurcation, resulting in a jump to large-amplitude limit cycles. As w increases further, these limit cycles lose stability via a secondary Hopf bifurcation at a value of w less than that in the case of $\tau = 1/64.4$. Decreasing w, one finds that the quasi-periodic solution loses stability via a reversed secondary Hopf bifurcation, resulting in a stable limit cycle. As w decreases further, the amplitude of the limit cycle decreases and loses stability via a cyclic-fold bifurcation, resulting in a jump to the trivial

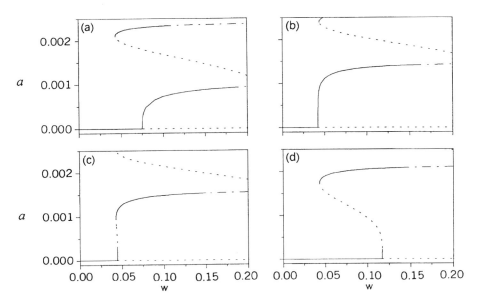

FIGURE 7.5. Nonlinear response curves for a six-term approximation when (a) $\tau = 1/60$, (b) $\tau = 1/64.4$, (c) $\tau = 1/66$, and (d) $\tau = 1/75$. Solid lines denote stable trivial or periodic solutions, dotted lines denote unstable trivial or periodic solutions, and dotted-dashed lines denote the region of quasi-periodic and chaotic solutions.

solution for a value of w less than that in the case of increasing w. Decreasing τ to $1/75$, we find that the interval between the two cyclic-fold bifurcations increases and the amplitude of the limit cycle resulting from the jump up also increases. Therefore, the larger the rotational speed is (i.e., the larger τ^{-1} is), the smaller is the value of w at which the secondary Hopf bifurcation occurs.

In Figure 7.6 we show variation of the amplitude a of the fundamental harmonic with $1/\tau$ for the two widths of cut 0.1 and 0.05. Again all Hopf bifurcations are supercritical. For some rotational speeds multiple solutions coexist, and hence the response depends on the initial conditions. These multiple solutions may correspond to stable cutting and one or more large-amplitude chatter, thereby yielding hysteresis and subcritical instabilities. As the rotational speed decreases, every Hopf bifurcation leads to the creation of a small-amplitude limit cycle that undergoes a cyclic-fold bifurcation, leading to a large-amplitude limit cycle. Again, as the rotational speed increases, every Hopf bifurcation leads to a small-amplitude limit cycle that smoothly grows and may undergo a secondary Hopf bifurcation. The interval between the Hopf bifurcations increases as the rotational speed or the width of cut increases.

7.5.4 Route to Chaos

Next we investigate the behavior of the motions for values of w exceeding that corresponding to the secondary Hopf bifurcation SH by using a numerical integration of the governing equations (4) for a time long enough for the transients to die out. To characterize the motions, we use four tools, namely phase portraits, time traces, Poincaré sections, and power spectra. For the Poincaré sections we collected all the points of intersection of the trajectory with the surface of section $x(t - \tau) = 0$ when $\dot{x}(t - \tau) > 0$.

In Figure 7.7 we show characteristics of the motion when $w = 0.136$. As expected, the motion is periodic: The phase portrait is a closed curve, the Poincaré section consists of a finite number of points (one point), and the power spectrum consists of a fundamental frequency and its even and odd harmonics, including a constant component (usually referred to as dc component). The magnitudes of the peaks are in good agreement with those obtained by using the method of harmonic balance.

As w increases, the limit cycle increases in size, deforms, and loses stability via a secondary Hopf bifurcation SH with a pair of complex conjugate eigenvalues γ crossing transversely from the left to the right half-plane. At $w \simeq 0.141$ the pair of eigenvalues is $\gamma = 0.1166 \pm 267.372i$, the chatter frequency $\omega_1 = 1391.07$, and the bifurcated solution can be characterized by two independent frequencies ω_1 and ω_2, which are, in general, incommensurate (i.e., ω_2/ω_1, is not a rational number). Characteristics of this motion are shown in Figure 7.8. Such a motion is called two-period quasi-periodic or two torus. The time trace suggests that the motion contains two periods and that the power spectrum clearly contains another frequency besides the original chatter

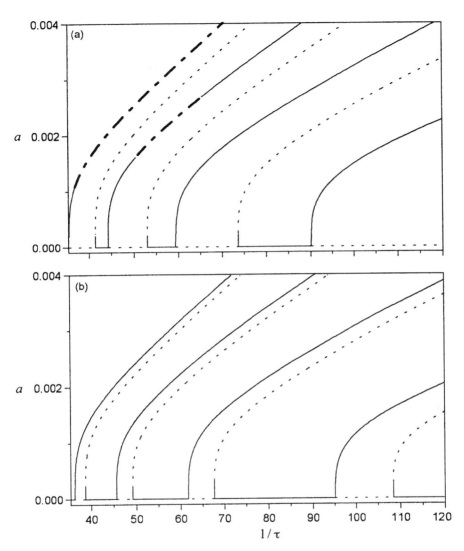

FIGURE 7.6. Nonlinear response curves for a six-term approximation when (*a*) $w = 0.1$ and (*b*) $w = 0.05$. Solid lines denote stable trivial or periodic solutions, dotted lines denote unstable trivial or periodic solutions, and dotted-dashed lines denote the region of quasi-periodic and chaotic solutions.

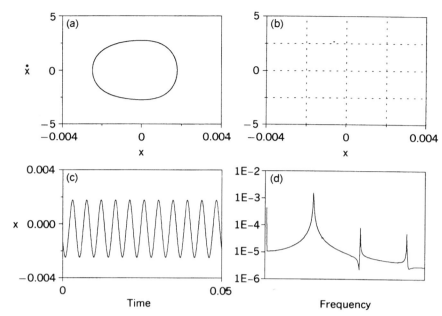

FIGURE 7.7. The phase portrait (*a*), the Poincaré section (*b*), the time history (*c*), and the FFT (*d*) of the attractor obtained when $\tau = 1/75$ and $w = 0.136$.

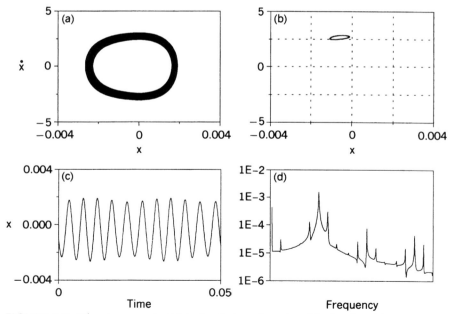

FIGURE 7.8. The phase portrait (*a*), the Poincaré section (*b*), the time history (*c*), and the FFT (*d*) of the attractor obtained when $\tau = 1/75$ and $w = 0.141$.

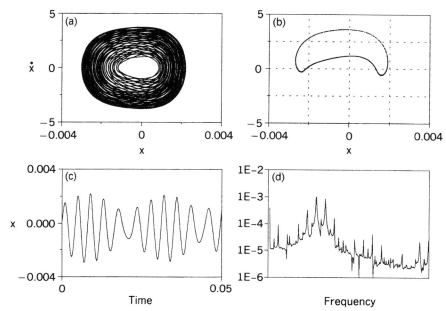

FIGURE 7.9. The phase portrait (*a*), the Poincaré section (*b*), the time history (*c*), and the FFT (*d*) of the attractor obtained when $\tau = 1/75$ and $w = 0.230$.

frequency. The ratio of the chatter frequency $\omega_1 = 1391.07$ to the modulated frequency $\omega_2 = 267.37$ is 5.203 approximately. As indicated by the power spectrum, the two peaks on both sides of the fundamental frequency component are about $0.4\omega_1$ apart. The phase portrait seems not to close on itself, suggesting a nonperiodic motion. The collected points on the one-sided Poincaré section fill up uniformly and densely a closed curve, confirming the two-period quasi-periodic nature of the motion. Such a motion can be calculated by using a spectral analysis (Nayfeh and Balachandran, 1995).

As w is increased substantially above 0.141, the Poincaré section gets distorted, and the collected points on the surface of section cover the closed curve nonuniformly, as shown in Figure 7.9 when $w = 0.230$. The power spectrum contains many more frequencies. As w increases further, the torus grows and doubles, as shown in Figure 7.10 for $w = 0.240$. The postbifurcation state is a new torus that forms two loops around the original torus. The power spectrum of the doubled torus has many more frequencies than that of the torus in Figure 7.9. Instead of undergoing a complete cascade of period-doubling bifurcations, the doubled torus deforms into a wrinkled one, followed by a fractal torus, and finally a destruction of the torus and the emergence of a chaotic attractor, as shown in the Poincaré sections Figures 7.11 and 7.12. This transition to chaos through a two-period quasi-periodic attractor is often described as chaos via torus breakdown (Nayfeh and Balachandran, 1995).

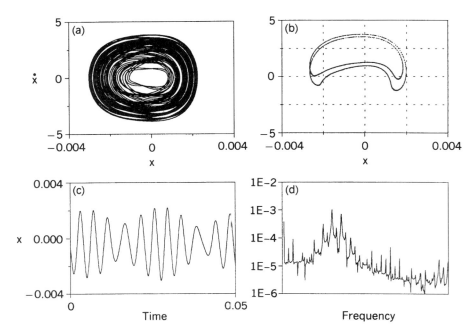

FIGURE 7.10. The phase portrait (a), the Poincaré section (b), the time history (c), and the FFT (d) of the attractor obtained when $\tau = 1/75$ and $w = 0.240$.

7.6   CONCLUSION

A brief account of the application of the methods of multiple scales, harmonic balance, and Floquet theory to tool chatter dynamics is described by using a single-degree-of-freedom model that accounts for nonlinear stiffness of the tool and tool holder and linear and nonlinear regenerative effects in the cutting force. The results show that a two-term harmonic approximation might lead to quantitatively as well as qualitatively erroneous results. On the other hand, a six-term harmonic approximation leads to results that are in full agreement with digital simulations of the model equation. For a given rotational speed, as the width of cut increases, the response remains trivial until a critical value is reached. Beyond this value a supercritical Hopf bifurcation occurs, leading to a loss of stability of the trivial solution and the creation of a small limit cycle. Depending on the rotational speed there are two scenarios. First, the limit cycle may grow smoothly and then lose stability via a secondary Hopf bifurcation, resulting in the creation of a quasi-periodic solution that bifurcates into chaos via torus breakdown. Second, the limit cycle created as a result of the Hopf bifurcation quickly loses stability via a cyclic-fold bifurcation that results in a jump to a large-amplitude limit cycle, which in turn loses stability

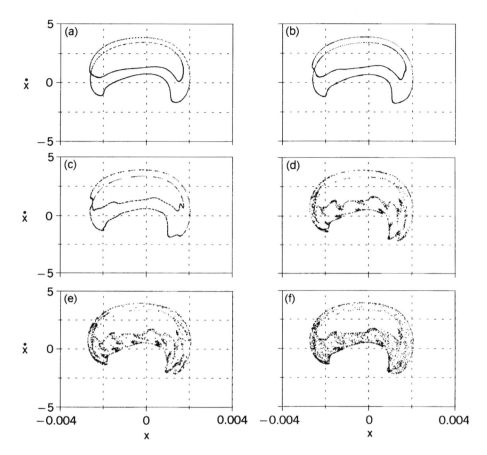

FIGURE 7.11. Poincaré sections of the attractors obtained when $\tau = 1/75$ and $w = (a)$ 0.250, (b) 0.255, (c) 0.260, (d) 0.265, (e) 0.266, and (f) 0.267.

via a secondary Hopf bifurcation, leading to a quasi-periodic solution. Then the quasi-periodic solution breaks down, resulting in a chaotic solution. The cyclic-fold bifurcations result in hysteresis and subcritical instabilities that lower the effective widths of cut for stable cutting.

ACKNOWLEDGMENTS

This work was supported by the National Science Foundation under Grant No. CMS-9423774.

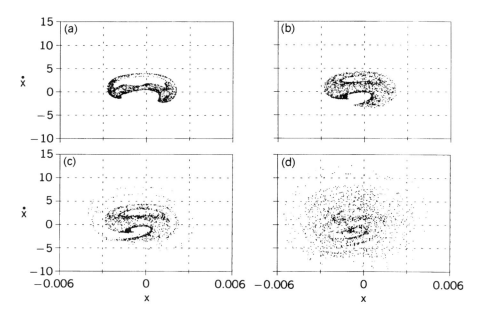

FIGURE 7.12. Poincaré sections of the attractors obtained when $\tau = 1/75$ and $w = (a)$ 0.27, (b) 0.30, (c) 0.32, and (d) 0.35.

REFERENCES

Arnold, R. N. 1946. The mechanism of tool vibration in the cutting of steel. *Proc. Inst. Mech. Eng.* **154**:261–284.

Grabec, I. 1986. Chaos generated by the cutting process. *Phys. Lett. A* **117**:384–386.

Grabec, I. 1988. Chaotic dynamics of the cutting process. *Int. J. Mach. Tools Manuf.* **28**:19–32.

Hana, N. H., and Tobias, S. A. 1974. A theory of nonlinear regenerative chatter. *ASME J. Eng. Ind.* **96**:247–255.

Koeingsberger, F., and Tlusty, J. 1970. *Machine Tool Structures*, vol. 1. Pergamon Press, Manchester.

Lin, J. S., and Weng, C. I. 1990. A nonlinear dynamic model of cutting. *Int. J. Mach. Tools Manuf.* **30**:53–64.

Lin, J. S., and Weng, C. I. 1991. Nonlinear dynamics of the cutting process. *Int. J. Mach. Tools Manuf.* **33**:645–657.

Minis, I., Magrab, E., and Pandelidis, I. 1990. Improved methods for the prediction of chatter in turning. III: A generalized linear theory. *ASME J. Eng. Ind.* **112**:28–35.

Moon, F. C. 1994. Chaotic dynamics and fractals in material removal processes. In Thompson, J. M. T., and Bishop, S. R., eds., *Nonlinearity and Chaos in Engineering Dynamics*. Wiley, New York.

Nayfeh, A. H. 1973. *Perturbation Methods*. Wiley, New York.

Nayfeh, A. H. 1981. *Introduction to Perturbation Techniques.* Wiley, New York.

Nayfeh, A. H., and Balachandran, B. 1995. *Applied Nonlinear Dynamics.* Wiley, New York.

Nayfeh, A. H., and Mook, D. T. 1979. *Nonlinear Oscillations.* Wiley, New York.

Nayfeh, A. H., Chin, C.-M., and Pratt, J. 1997. Perturbation methods in nonlinear dynamics — Applications to machining dynamics. *J. Manuf. Sci. Eng.,* forthcoming.

Pratt, J., and Nayfeh, A. H. 1996. Experimental stability of a time-delay system. *Proc. 37th AIAA/ASME/ASCE/AHS/ACS Structures, Structural Dynamics, and Materials Conf.,* Salt Lake City, UT, April 15–17.

Tlusty, J., and Ismail, F. 1982. Basic nonlinearity in machining chatter. *CIRP Ann.* **30**:229–304.

Tobias, S. A. 1965. *Machine-Tool Vibration.* Wiley, New York.

Wu, D. W., and Liu, C. R. 1985. An analytical model of cutting dynamics. I: Model building. *ASME J. Eng. Ind.* **107**:107–111.

8

NONLINEAR DYNAMICS
AND SURFACE
INTEGRITY IN
MACHINING CERAMICS

GUANGMING ZHANG, STANLEY J. NG, and DUNG T. LE

8.1 INTRODUCTION

Advanced ceramics offer many desirable characteristics for industrial and commercial use in terms of their high-temperature tolerance, wear and abrasive resistance, and corrosion resistance. These intrinsic properties have led ceramics to be prime candidates for many applications ranging from engineering components to dental restorations. However, the machining of ceramics into practical forms presents a challenge because of the nonlinear dynamics involved in the material removal process inherent by its high hardness and high brittleness.

Manufacturing of ceramic parts requires dimensional and geometrical accuracy. Machining has played a key role in this respect to ensure the functionality of surfaces of ceramic parts, especially after sintering. However, due to their hard and brittle nature, ceramic materials are difficult to machine. The fracture-dominated material removal process in ceramics machining leaves cracks on and beneath the machined surfaces, forming surface and subsurfacing damage and leading to a shortened product life cycle. Moreover the high wear rate of cutting tools or grinding wheels during machining gives rise to high production cost. The nonlinear dynamics can be best described by the basic

Dynamics and Chaos in Manufacturing Processes, Edited by Francis C. Moon.
ISBN 0-471-15293-5 © 1998 John Wiley & Sons, Inc.

five stages of the material removal mechanism in advanced ceramic material, as illustrated in Figure 8.1.

To overcome these difficulties associated with ceramic machining, research has been extensive with focus on two areas: (1) searching innovative machining technologies that maintain the materials' special features while being cost-effective and (2) developing evaluation methodologies to assess the machining performance, especially to characterize the nonlinear dynamics of the material removal process. In the first area, new technologies to process ceramics have been developed. For example, ductile grinding of optical glass without cracks was reported by Evan and Marshall (1980), Inasaki (1987), and Konig et al. (1991), although the material removal rate had to be restricted at a very low level. High-speed grinding, electro-discharge machining, and laser-assisted machining have been exploited for machining ceramics with success, such as by Mazurkiewicz (1991), Nakagawa et al. (1985), and Zhang et al. (1994). However, the high investment costs associated with these machining techniques

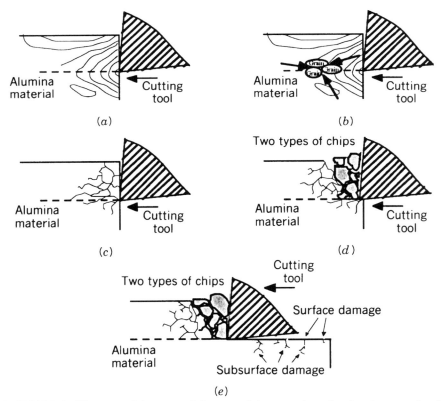

FIGURE 8.1. Five essential stages of the material removal mechanism in ceramics: (a) Dynamic loading and induced stress field; (b) crack initiation; (c) crack propagation; (d) chip formation; (e) formation of surface and subsurface damage.

limit their applications. In the second area where research on ceramic machining has been extensive, methods to characterize surface and subsurface structures present in the ceramics after machining have been developed by Lawn (1994), Slotwinski et al. (1993), Jahanmir et al. (1992), and Groenou et al. (1979). Ultrasonic measurements have been used to study the surface texture formed during machining by monitoring the amplitude of the generated surface waves. Indentation techniques have also been used to investigate the residual stresses in the machined ceramics. These trends in ceramic machining research show the importance of having nondestructive evaluation methods that are effective and sensitive in detecting the surface and subsurfacing damage formed during machining.

In this chapter we present our research on the study of nonlinear dynamics observed during the machining of ceramic materials. We will focus on both innovation in machining technology and development of nondestructive evaluation methods to assess machining performance. Three aspects of our work are presented in the following sections. They are (1) submerged precision machining and (2) scanning electron microscopy analysis. Aspect 1 relates to an innovative approach to machining, while aspects 2 relates to evaluation methodologies for surface characterization. These methods and techniques have been developed to achieve cost-efficient machining as well as high-quality surface finish in ceramic material.

8.2   SUBMERGED PRECISION MACHINING

We first present a new method of machining based on the stress-corrosion behavior of ceramic materials. This is a passive method for controlling the nonlinear dynamics; it utilizes a submerged machining environment to assist in the material removal process. Results from our research have indicated that reducing friction at contact surfaces and removing heat generated during machining, thereby changing the stress field distribution in the immediate cutting zone. The advantages of this process can be further enhanced through proper choice of industrial machining coolants, and chemical additives may provide a better control over strength and fracture toughness of ceramic material. It is assumed that the submersion environment will markedly improve surface quality and reduce subsurface damage.

8.2.1   Basic Methodology

The submersion method can be easily adapted to machine ceramics. The submerged machining apparatus is designed and constructed, as shown in Figure 8.2, so that the ceramic material to be machined is placed in a bath filled with cutting fluid. In particular, the device consists of three parts, namely a container for cutting fluid, a vice to hold the workpiece, and a base on which force sensors are mounted for the purpose of measuring the cutting force

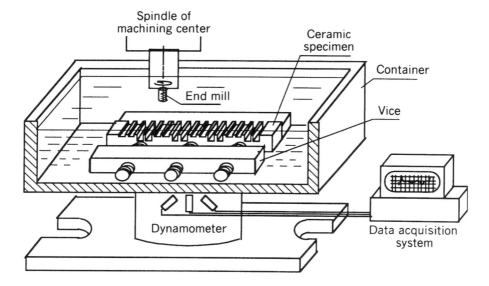

FIGURE 8.2. Submerged machining apparatus.

during machining. During machining tests the apparatus is fixed on the table of a CNC machining center. The workpiece is tightened in the vice before filling the bath with cutting fluid. By controlling the type of cutting fluid and machining parameters, namely the depth of cut, feed, cutting speed, and the temperature in the cutting zone, the tribological effects on the tool-workpiece and tool-chip interfaces are brought into play in the material removal process. It has been observed that the submerged machining environment promotes stress-corrosion cracking such as is induced by chemical reactions through high penetration of the cutting fluid. Experimental results strongly support the potential of using submerged machining to improve quality and efficiency of the machining operation.

8.2.2   A Case Study of Submerged Machining

To demonstrate the effectiveness of submerged machining, a case study is presented. The ceramic material used in the study is a machinable glass ceramic (MGC) from Corning. Its chemical formula is $K_2O-MgF_2-MgO-SiO_2$. It is a two-phase tetrasilicic mica glass-ceramic material and has a microstructure consisting of mica flakes of approximately 70 vol.% dispersed in a nonporous glass matrix. The hardness of this material is about 3.4 GPa. The cleavage fracture along the planes of mica flakes makes the material machinable. This MGC has been widely used as a new material in dental restorations. Figure 8.3 illustrates the specimen used in the machining tests. It is a 100 mm long bar with a rectangular cross section, each side equal to 12.7 mm. The material of the end mill, which has a diameter of 3.175 mm, was high-speed steel with hardness of about 18 GPa, which is five times as hard as the MGC material under the investigation.

The four parameters considered in this study are the type of cutting fluid used and the three machining parameters (depth of cut, feed rate, and cutting speed), whose settings are implemented through a numerically controlled (NC) program. A design of experimentation method, as shown in Figure 8.3 from Box et al. (1978), is used to set each of the four variables at two levels, namely, high and low. Table 8.1 lists the values of the high and low levels of the three machining parameters. For the cutting fluid a commercially available emulsifiable oil, called LS-A-14H, is used from Tower Oil and Technology Co. (1994). The emulsifiable concentrate is mixed with water in a ratio of 9:1. The pH ranges from 9.7 to 9.9. An NC program was prepared to carry out the machining tests, first at the dry environment and then at the submersion environment.

Assessment of experimental data obtained from the submerged machining investigation is based on the requirements set for the preparation of dental restorations using ceramic material. One of these requirements is reliability or durability of a restoration during its service. In fact any fabrication method will cause damage to the ceramic during the restoration preparation. It is believed that the smoother the surface produced, the less likelihood that the ceramic has been damaged sufficiently to compromise its clinical performance. In addition quantitative information on the degree of the damage is essential to the assessment of machining performance. Observations indicate that failure of most of the ceramic components occurring during service initiates at the surface cracks, which grow and penetrate, leading to unstable failure.

The assessment performed in this study consists of two aspects. The first aspect is to measure finish quality of machined surfaces using the traditional profilometer method. A surface profilometer, Perthometer-S5P (resolution 40×10^{-9} m), is used to make traces from each of the 16 machined surfaces, and values of roughness average, R_a, are obtained. They are listed in Figure 8.3. The second aspect is to quantify cracks formed on the machined surface through visualization of the surface texture formed during machining. To do so, an environmental scanning electron microscope is used to obtain high-magnification images of localized areas of surface texture, which are representative of the machined surface, as detailed by Goldstein et al. (1992). Figure 8.4 presents a set of electron micrographs taken from the 16 machined surfaces. From these images, reconstruction of the surface texture formed during machining is realized using the image-processing method discussed in the next section on scanning electron microscopy analysis. The two reconstructed 3-dimensional surface topographies shown in Figure 8.5 are taken from the two pictures marked as test 2 and test 10 in Figure 8.4. To quantify the damage induced under different machining conditions, a performance index is introduced to characterize the machined damage in terms of surface cracking (results are also shown in Figure 8.3). The performance index is called the

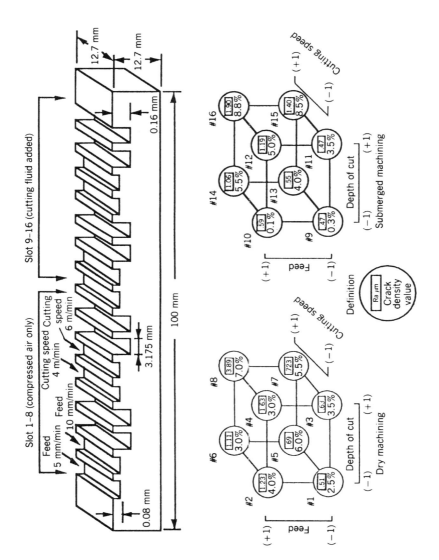

FIGURE 8.3. Ceramic specimen and settings of the machining conditions.

TABLE 8.1
Value Settings for Design of Experimentation

Parameters	Cutting Speed	Feed Rate	Depth of Cut	Environment
High (+1)	6 m/min	10 mm/min	0.16 mm	Submerged
Low (−1)	4 m/min	5 mm/min	0.08 mm	Dry

density of surface cracking, and it is defined as the percentage of the area occupied by cracks formed on a machined surface. As illustrated in Figure 8.6, two contour plots are taken from the two reconstructed surface topographies for evaluating the density of surface cracking. As Figure 8.6 shows, the surface cracks have a cone-shaped geometry. The crack density depends on the level of depth with which a contour plot is being taken. Figure 8.5 gives the two contours shown in Figure 8.6 at a level of depth marked as CP − 1.7σ, where parameter σ represents the standard deviation of the height variation about the center plane, CP. To further clarify the two parameters CP and σ, formulas for determining them during the evaluation are

$$CP = \sum_{i=1}^{n} \sum_{j=1}^{n} \frac{y_{ij}}{n \times n}, \tag{1}$$

$$\sigma = \sqrt{\sum_{i=1}^{n} \sum_{j=1}^{n} \frac{(y_{ij} - CP)^2}{n \times n}}, \tag{2}$$

where y_{ij} represents the height of an individual point with respect to a reference plane and n represents the number of points taken on each side of the squared area for the evaluation. Results obtained from evaluating the density of surface cracking for the 16 machined surfaces at several levels of depth are plotted in Figures 8.6a and 8.6b for the dry and submerged machining environment, respectively. The data listed in Figure 8.3 represent the crack density values evaluated at the level of depth set at 1.7σ below CP.

8.2.3 Analysis of Results

Data obtained from a factorial design, such as the one used in this study, offers an unique opportunity to derive empirical models capable of describing effects of the four parameters on the machining performances of dry and submerged environments.

Main Effects of the Four Machining Parameters

Using the algorithm, proposed by Box et al. (1978), averaging the eight pairs of roughness mesurements for each of the four machining parameters provides

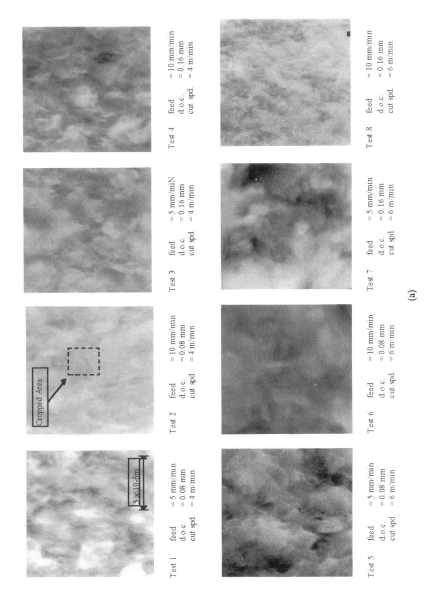

Test 1 feed   = 5 mm/min
         d.o.c.   = 0.08 mm
         cut spd.   = 4 m/min

Test 2 feed   = 10 mm/min
         d.o.c.   = 0.08 mm
         cut spd.   = 4 m/min

Test 3 feed   = 5 mm/miN
         d.o.c.   = 0.16 mm
         cut spd.   = 4 m/min

Test 4 feed   = 10 mm/min
         d.o.c.   = 0.16 mm
         cut spd.   = 4 m/min

Test 5 feed   = 5 mm/min
         d.o.c.   = 0.08 mm
         cut spd.   = 6 m/min

Test 6 feed   = 10 mm/min
         d.o.c.   = 0.08 mm
         cut spd.   = 6 m/min

Test 7 feed   = 5 mm/min
         d.o.c.   = 0.16 mm
         cut spd.   = 6 m/min

Test 8 feed   = 10 mm/min
         d.o.c.   = 0.16 mm
         cut spd.   = 6 m/min

Cropped Area

5×10^{-6}m

(a)

222

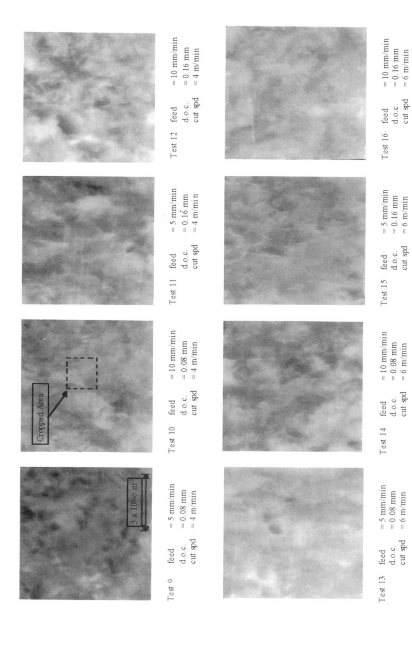

Test 9   feed   = 5 mm/min
          d.o.c.   = 0.08 mm
          cut spd = 4 m/min

Test 10   feed   = 10 mm/min
          d.o.c.   = 0.08 mm
          cut spd = 4 m/min

Test 11   feed   = 5 mm/min
          d.o.c.   = 0.16 mm
          cut spd = 4 m/min

Test 12   feed   = 10 mm/min
          d.o.c.   = 0.16 mm
          cut spd = 4 m/min

Test 13   feed   = 5 mm/min
          d.o.c.   = 0.08 mm
          cut spd = 6 m/min

Test 14   feed   = 10 mm/min
          d.o.c.   = 0.08 mm
          cut spd = 6 m/min

Test 15   feed   = 5 mm/min
          d.o.c.   = 0.16 mm
          cut spd = 6 m/min

Test 16   feed   = 10 mm/min
          d.o.c.   = 0.16 mm
          cut spd = 6 m/min

(b)

FIGURE 8.4. SEM micrographs of the eight machined surfaces: (*a*) Dry machining environment; (*b*) submerged machining environment.

223

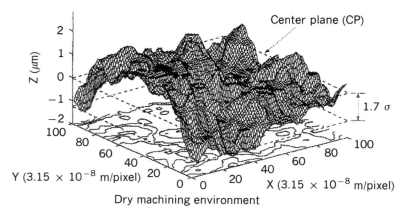

(a) Test 2: Feed = 10 mm/min (+1), Depth of cut = 0.08 mm (−1),
Cutting speed = 4 m/min (−1)

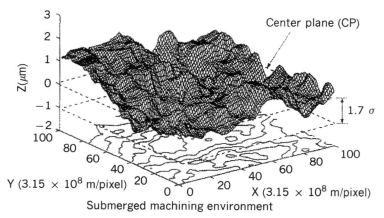

(b) Test 10: Feed = 10 mm/min (+1), Depth of cut = 0.08 mm (−1),
Cutting speed = 4 m/min (−1)

FIGURE 8.5. Two reconstructed surface topographies.

an estimate of the main of each of the four parameters on finish quality in terms of surface roughness. For example, the main effect of the machining environment can be identified by

$$R_a^- = \frac{1}{2}\left[\frac{0.47+0.59+0.47+1.19+0.55+1.05+1.40+1.90}{8}\right.$$
$$\left. - \frac{0.51+1.23+0.63+1.63+0.69+1.11+0.72+3.89}{8}\right] = -0.18 \quad (\mu m).$$

$$(3)$$

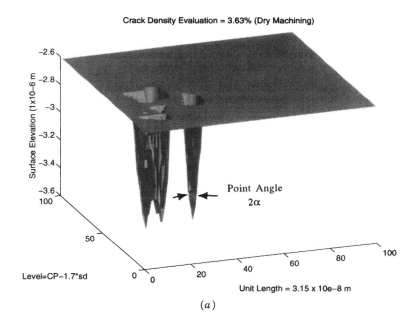

Crack Density Evaluation = 3.63% (Dry Machining)

Point Angle
2α

Surface Elevation (1x10–6 m)

Level=CP–1.7*sd

Unit Length = 3.15 x 10e–8 m

(a)

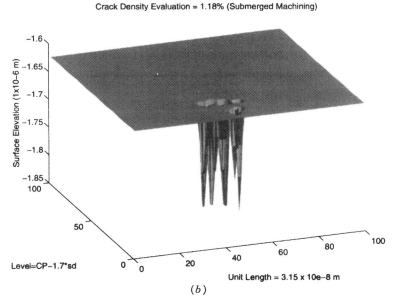

Crack Density Evaluation = 1.18% (Submerged Machining)

Surface Elevation (1x10–6 m)

Level=CP–1.7*sd

Unit Length = 3.15 x 10e–8 m

(b)

FIGURE 8.6. Contour plots for the evaluation of surface cracking obtained from stereophotogrammetric examination of two machined surfaces: (a) Dry environment; (b) submerged environment.

225

The three main effects of feed, depth of cut, and cutting speed are obtained in a similar manner. An empirical model to describe the main effects of these four parameters on finish quality is represented by

$$R_a^- = 1.13 + 0.45(\text{feed}) + 0.35(\text{depth of cut})$$

$$+ 0.28(\text{cutting speed}) - 0.18(\text{machining environment})$$

$$+ (\text{combinatorial effects}) \quad (\mu\text{m}). \tag{4}$$

Note that the first term in Equation (4), $1.13\,\mu\text{m}$, is the grand average, or the average of the $16\,R_a$ measurements; it indicates the order of surface finish under this investigation, which is about $1\,\mu\text{m}$. In Equation (4) the numerical values of the four parameters used for prediction have to be either $+1$, representing a high-level setting, or -1, representing a low-level setting. For the qualitative parameter representing the machining environment, -1 means dry machining and $+$submerged machining. For example, replacing all the parameters by -1 in Equation (3) corresponds to a combination of smaller depth of cut, smaller feed, lower cuting speed, and a dry machining environment.

Examining Equation (4), the four values associated with the four machining parameters characterize their main effects on the surface finish. The three positive values indicate that large R_a values, or a degrading trend of surface finish, can be anticipated when feed, depth of cut and cutting speed are set at the high level. On the other hand, the negative value $-0.18\,\mu\text{m}$ is associated with the submersion machining environment parameter, revealing the benefit of improved finish quality. The ratio of 0.18 to 1.13 shows a 16% of reduction from the mean level of roughness average when the submersion is being used to machine the Corning-MGC material.

Interaction Effects between and/or among the Four Parameters

The benefit of using submerged machining can be further exploited when the experimental results are analyzed in two separated sets, one with the dry machining environment and the other with the submerged machining environment. For each data set, the grand average, main effects, and combinatorial effects among feed, depth of cut, and cutting speed are calculated to obtain the two empirical models. It is important to note that these combinational effects characterize the nonlinear behavior of the material removal in the surface texture formation.

For the dry machining environment,

$$R_a^- = 1.30 + 0.66(\text{feed}) + 0.42(\text{depth of cut}) + 0.30(\text{cutting speed})$$

$$+ 0.38(\text{feed} \times \text{depth of cut}) + 0.23(\text{feed} \times \text{cutting speed})$$

$$+ 0.29(\text{depth of cut} \times \text{cutting speed})$$

$$+ 0.31(\text{feed} \times \text{depth of cut} \times \text{cutting speed}) \quad (\mu\text{m}). \tag{5}$$

For the submerged machining environment,

$$R_a^- = 0.95 + 0.23(\text{feed}) + 0.29(\text{depth of cut}) + 0.27(\text{cutting speed})$$
$$+ 0.08(\text{feed} \times \text{depth of cut}) + 0.02(\text{feed} \times \text{cutting speed})$$
$$+ 0.14(\text{depth of cut} \times \text{cutting speed})$$
$$- 0.08(\text{feed} \times \text{depth of cut} \times \text{cutting speed}) \quad (\mu\text{m}). \tag{6}$$

Examining the eight values representing the grand average, the main and combinational effects in Equations (5) and (6), the following important observations are made:

1. The numerical values in Equation (6), or associated with the submersion, are significantly smaller than those in Equation (5), indicating that a better surface finish codition can always be expected when the submersion is applied. This can be further confirmed by examining the data presented in the two cubics shown in Figure 8.3. For the identical machining parameter settings, the roughness value for test 1 is 0.51 μm for the dry machining, larger than 0.49 μm for test 9 under the submerged machining, 1.23 μm for test 2 which is larger than 0.47 μm for test 10, and so on.

2. The improved finish quality due to submerged machining is unevenly distributed among the eight effects estimated from the experimental data. Figure 8.7 presents a comprehensive picture of such a distribution. The

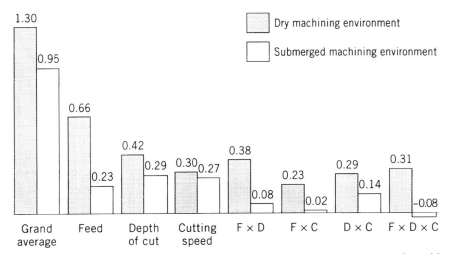

FIGURE 8.7. Comparison of estimated effects between the dry and submerged machining environments.

most significant reductions are associated with the combinational effect between feed and depth of cut, the combinational effect between feed and cutting speed, and the combinational effect among feed, depth of cut, and cutting speed. Statistically these reductions almost eliminate the degrading effect on the finish quality. Significant reductions include the grand average, the main effect of feed, the main effect of depth of cut, and the combinational effect of depth of cut and cutting speed. The lowest reduction, which is 10%, is associated with the main effect of cutting speed. Statistically the following formula gives the relationship between the roughness average of a machined surface and the three machining parameters under the submerged machining:

$$R_a^- = 0.95 + 0.23(\text{feed}) + 0.29(\text{depth of cut}) + 0.27(\text{cutting speed})$$

$$+ 0.14(\text{depth of cut} \times \text{cutting speed}) \quad (\mu\text{m}). \tag{7}$$

Equation (7) expresses the presence of stress corrosion cracking in the ceramic material during material removal. Because of the brittle fracture, the mechanics of material removal during the machining of ceramics is dominated by nonlinear dynamics.

3. The interplay between depth of cut and cutting speed, combining low cutting speed and low depth of cut, may result in finish quality that is compatible with the finish quality obtainable under the combination of low cutting speed and high depth of cut. This observation has practical interest, for, when submersion is applied, a combination of low cutting speed and large depth of cut can offer a higher material removal rate while keeping a satisfactory finish to the machined surface.

Thus experimental results have confirmed the validity of using submerged machining as an innovative machining technique to control the surface finish of ceramics. In the next section an effort is made to explain the technique of using an environmental scanning electron microscope (ESEM) to obtain high-quality, high-magnification images of machined ceramic surfaces for the purpose of surface finish assessment. The use of an ESEM, together with computer graphics, presents an improvement over the traditional means for measuring surface roughness.

8.3   ASSESSMENT OF SURFACE INTEGRITY USING SEM

We have developed a method for assessing surface integrity by scanning electron microscopy (SEM). A computer-based system has been set up to evaluate the condition of the machined surface. As illustrated in Figure 8.8, a ceramic specimen with a machined surface is first examined under environmen-

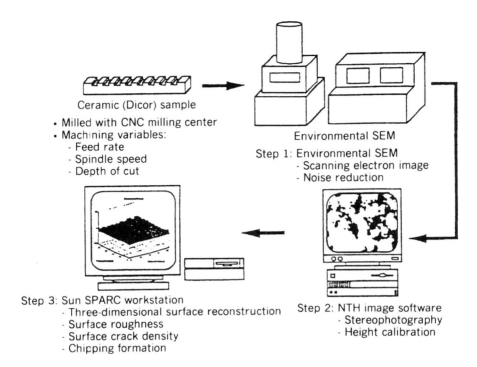

Ceramic (Dicor) sample
- Milled with CNC milling center
- Machining variables:
  - Feed rate
  - Spindle speed
  - Depth of cut

Environmental SEM

Step 1: Environmental SEM
  - Scanning electron image
  - Noise reduction

Step 3: Sun SPARC workstation
  - Three-dimensional surface reconstruction
  - Surface roughness
  - Surface crack density
  - Chipping formation

Step 2: NTH image software
  - Stereophotography
  - Height calibration

FIGURE 8.8. Computer-based system for surface characterization.

tal scanning electron microscopy (ESEM). SEM micrographs of representative locations on the machined surface are displayed on the screen, inspected, and digitized. Corresponding image files are then stored on computer. Numerical data in the image files are the recorded intensity of electron reflection on the surface texture. To obtain the height information on the surface images, a grayscale-height calibration process is performed using the stereo-pair method, as reported by Slotwinski (1993). When the SEM image data files are transformed to the height variation data files, three-dimensional topographies in micro-scale can be reconstructed using computer graphics.

SEM photography is a well-known method for obtaining high-resolution images with a great depth of field of machined surfaces. Through the use of advanced computer image-processing techniques, qualitative information regarding surface morphology can be now extracted from the SEM micrographs and be evaluated quantitatively to describe surface characterists such as surface roughness (R_a), residual stress distribution, and the density of surface cracking in terms of the percentage of surface area. The use of this nondestructive evaluation (NDE) technique can provide important feedbacks to the user for optimization of machining parameters.

8.3.1   Applications: Study of Surface Texture and Surface Roughness

We will demonstrate the procedure by using the SEM-stereophotography method to characterize surface texture and a machinable dental ceramic, Dicor-MGC polycrystal, to prepare the specimens with machined surfaces. This is the same material used in the previous case study. According to Groenou et al. (1979), this material matches the physical properties of human enamel in terms of translucency, thermal conductive, density, and hardness.

The preparation of specimens with machined surfaces is illustrated in Figure 8.9. An end mill with a diameter of 3.175 mm cuts two blocks of $12.7 \times 12.7 \times 44.5\,mm^3$. On each of the two blocks, four slots are machined under four different conditions. As reported in Zhang et al. (1995), the four different machining conditions for slots 1–4 are combinations of two feed rate

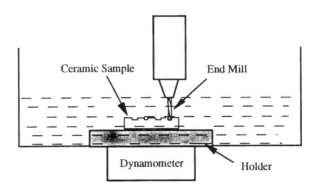

Test	Feed (mm/min)	Speed (r.p.m.)	Depth (mm)
1	5	600	0.5
2	5	900	0.5
3	10	600	0.5
4	10	900	0.5

Test	Feed (mm/min)	Speed (r.p.m.)	Depth (mm)
5	5	600	0.25
6	5	900	0.25
7	10	600	0.25
8	10	900	0.25

FIGURE 8.9. Preparation of specimens with machined surfaces.

settings (5 and 10 mm/min) and two spindle speed settings (600 and 900 rpm) with the depth of cut set at 0.5 mm. When preparing slot 5–8, the feed rate and spindle speed settings remain unchanged, but the depth of cut is set at 0.25 mm.

The specimens after machining are examined using ElectroScan E-3 environmental scanning electron microscope. Figure 8.10 presents eight representative SEM micrographs obtained through image processing. Each of these eight graphs contains 125 pixels by 125 pixels, representing a scanning area of $0.10 \times 0.10\,mm^2$. The image data are then transformed into height variation data using a defined calibration relation between the intensity of electron reflection and the micro-scale surface graduation. Figure 8.11 presents two surface topographs reconstructed in the three-dimensional space and their respective contour maps after the data transformation.

Examining Figure 8.11, the two reconstructed surface topographies represent visualization of machined surfaces. The surface texture shown in Figure 8.11a characterizes the machined surface on slot 1, and the surface texture shown in Figure 8.11b characterizes the machined surface on slot 4. A comparison of the two surface topographies gives a vivid picture of how the two surface textures are different from each other. The surface condition on slot 4 is much rougher than that on slot 1 because it has higher peaks and deeper valleys.

Note the scale used in the visualization. In the marked X- and Y-axes, each pixel represents $0.75\,\mu m$, and the indicated 80 pixels represent a length of $60\,\mu m$. The unit of the vertical axis is micrometer. To quantify the surface texture, contour plots are taken at six levels on the vertical direction. They are $2.0\,\mu m$, $1.5\,\mu m$, $1.0\,\mu m$, $0.5\,\mu m$, $0\,\mu m$, and $-1.0\,\mu m$ with the level of $0\,\mu m$ representing the reference plane used for the R_a evaluation as illustrated in Figure 8.5. Figure 8.7 presents two contour maps. Figure 8.11a gives the contour map representing slot 1. The contour map is an assembly of the six contour plots at the six indicated height levels taken from the surface topography shown in Figure 8.5a. Similarly Figure 8.11b gives the contour map representing slot 4. On these contour maps the grain size of $2\,\mu m$ is also indicated as reference for examination.

Examining the two contour maps, several important observations can be made. The first observation is the presence of isolated islands formed by clustered contour lines on the contour maps. These isolated islands represent valleys formed on the surface texture. The size of these isolated islands characterizes the geometric shape of their corresponding valleys. It is interesting to note that the smallest valley on both maps is about $2 \times 2\,\mu m$, which is about the grain size of the Dicor material used in this study. This observation suggests the existence of pullouts of a single grain during machining. It is well understood that grain boundaries in Dicor material represent a weak link when subjected to a tensile stress field due to the presence of second phase. The development of micro-cracking on the grain boundary enforces an entire grain to dislodge from the surface being machined. Valleys or cavities on the surface texture are formed as a result of these pullouts. The second observation is that

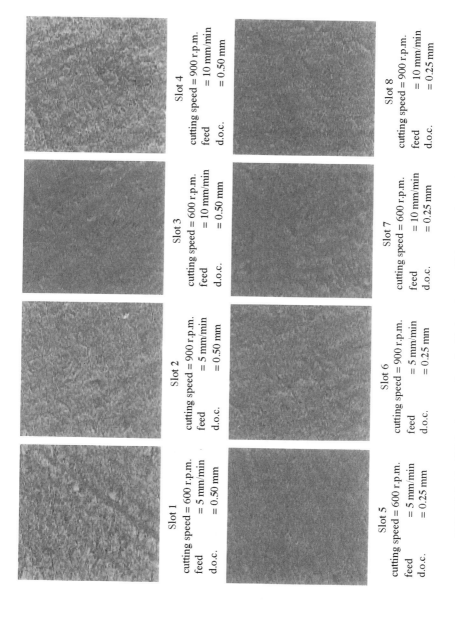

FIGURE 8.10. SEM micrographs of milled surfaces of Dicor specimen (area $= 0.10 \times 1.10$ mm).

Slot 1
cutting speed = 600 r.p.m.
feed   = 5 mm/min
d.o.c.   = 0.50 mm

Slot 2
cutting speed = 900 r.p.m.
feed   = 5 mm/min
d.o.c.   = 0.50 mm

Slot 3
cutting speed = 600 r.p.m.
feed   = 10 mm/min
d.o.c.   = 0.50 mm

Slot 4
cutting speed = 900 r.p.m.
feed   = 10 mm/min
d.o.c.   = 0.50 mm

Slot 5
cutting speed = 600 r.p.m.
feed   = 5 mm/min
d.o.c.   = 0.25 mm

Slot 6
cutting speed = 900 r.p.m.
feed   = 5 mm/min
d.o.c.   = 0.25 mm

Slot 7
cutting speed = 600 r.p.m.
feed   = 10 mm/min
d.o.c.   = 0.25 mm

Slot 8
cutting speed = 900 r.p.m.
feed   = 10 mm/min
d.o.c.   = 0.25 mm

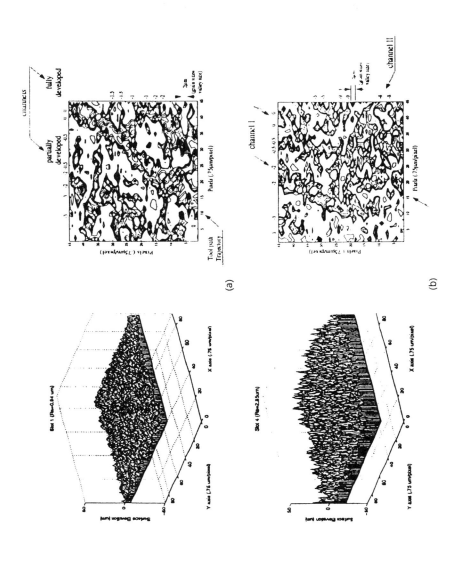

FIGURE 8.11. Reconstructed surface topography and contour map: (*a*) Machined surface, slot 1; (*b*) machined surface, slot 4.

233

most of the islands shown in Figure 8.7a are larger than the $2 \times 2\,\mu m$ islands. The average size of these islands is somewhere between $10 \times 5\,\mu m$ and $10 \times 10\,\mu m$. The $10\,\mu m$ is only one dimension of a micra flake. We note that the Dicor material used in this study contained mica flakes in a second phase. The additional mica flakes improve machinability as in the free machining of steel during which inclusions are added. This observation indicates that the easy cleavage properties of mica flakes create the weak boundaries that promote the micro-cracking propagation. By closely examining the large islands shown in Figure 8.7a, small islands can be identified, indicating the pullouts of single grains and the progressive development of micro-cracking between neighboring grain boundaries.

The third observation is the presence of cavity channels. Some of them are fully developed and some are partially developed, as illustrated in Figure 8.7a. The formation of these cavity channels is closely related to the internal stress field developed on/near the machined surface during machining. The most noticeable characteristic of the cavity channels is that their orientation follows the tool-path trajectory. This suggests that the tensile stress induced by machining is along the cutting speed direction. Under high internal stresses, micro-cracks merge along the tensile stress direction while releasing the strain energy accumulated under the builup of hydrostatic stresses. As long as a sufficient amount of the released energy is available, cavity channels will be formed during the material removal process. Otherwise, partially completed channels will be formed.

Traditionally surface texture formed during machining is examined using profilometers. By taking profiles, surface characterization indexes, such as roughness average and peak-to-valley, are evaluated from the measured heights along the profiles. Figure 8.12 presents four profiles that were constructed using the data stored in the height variation data file. The two profiles with $R_a = 0.83$ and $0.88\,\mu m$ in Figure 8.12a represent the surface roughness condition of slot 1, and the two profiles with $R_a = 3.25$ and $2.97\,\mu m$ in Figure 8.12b are representations of slot 4. Note the "cutoff length," which is $0.68\,\mu m$, used in the roughness average evaluation. It is $0.68\,\mu m$, representing a total of 90 data (pixels) used in the evaluation. Table 8.2 lists the roughness average values evaluated from the eight machined surfaces. The smallest R_a value is $0.21\,\mu m$ associated with slot 8. The corresponding machining conditions are feed rate: 5 mm/min; spindle speed: 600 rpm; and axial depth of cut: 0.25 mm. Numerical values of the standard deviation are also presented in Table 8.2, indicating the variation range of the R_a evaluation. In general, the standard deviation remains at $0.03\,\mu m$. As the R_a mean value increases, the standard deviation value increases accordingly.

8.4   CONCLUSION

This chapter has described the nonlinear dynamics observed during the

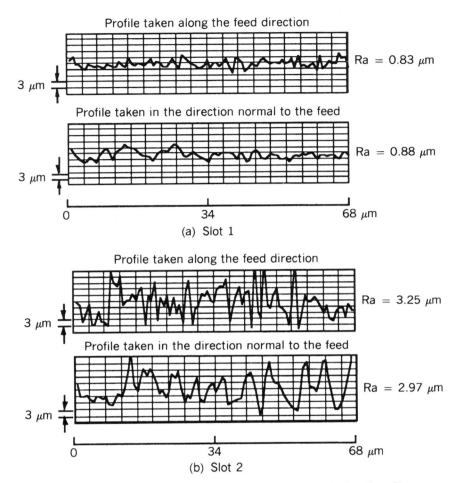

FIGURE 8.12. Profiles reconstructed from height variation data files.

machining of advanced ceramic materials. Brittle fracture dominates chip formation during the material removal process, so machining mechanics is shifted from shear deformation to micro-scale crack initiation, crack propagation, and macro-scale fracture. As this chapter has shown, investigation on the nonlinear dynamics has been intensive and significant progress has been made.

8.4.1 Submerged Machining

Significant findings have come from an experimental investigation that explored the advantages of using submerged machining to improve the surface integrity of ceramic parts. Surface finish and density of surface cracking were used as the two performance indexes for the performance evaluation. There

TABLE 8.2
Results of R_a

Slot Number	1	2	3	4	5	6	7	8
R_a (μm) 0.94	0.80	2.85	1.73	1.43	0.53	1.21	0.21	
Standard deviation (μm)	0.03	0.03	0.13	0.15	0.11	0.04	0.09	0.03

were three important observations:

1. Better performance in terms of improving surface finish and reducing the density of surface cracking. The chemomechanical effects contributed by stress-corrosion cracking assist in the chip formation process and increase the efficiency of material removal during machining.

2. The machining parameters had to be carefully set up in order to maximize the potential to achieve a smooth surface with a low surface cracking. It is recommended that low cutting speed and low depth of cut be used. The interaction between the depth of cut and cutting speed deserves special attention, due to the nonlinear behavior displayed in the interaction between the two parameters.

3. Fracture analysis is an area of investigation that needs further exploration for its potential in chemical-assisted machining of ceramic materials.

8.4.2   Assessment of Surface Integrity

To understand the nonlinear dynamics occurring during the machining of ceramic materials, quantitative measurements are critical. In this chapter a nondestructive evaluation approach was demonstrated. The approach integrated SEM analysis, signal processing, and computer graphics to visualize the surface texture formed during machining in micro-scale. Two performance indexes were introduced to quantify the effects of the nonlinear dynamics on surface integrity.

1. The SEM-stereophotography method developed in this study has successfully detected micro-scale surface textures. The high resolution of the micrographs ensures detailed information on surface height variation in micro-scale at an accuracy conventional profilometry methods can never achieve.

2. Identification of surface cavities has provided evidence of material removal mechanisms present during machining. Pullouts of individual grains due to dislodgment from the neighboring contact area and cavities

formed through the progressive development of grain boundary micro-cracking are two among the possible material removal mechanisms.

3. Factors including material microstructure and three machining parameters, namely feed rate, depth of cut, and spindle speed, were studied with respect to their influence on the surface texture formation. To quantify these effects, a new parameter, denoted as cavity density, was introduced. It was found that feed rate and depth of cut are the most influential factors in controlling the cavity density and that the spindle speed has the least influence because of the conflicting effects on the crack toughness of ceramic material contributed by the loading rate and temperature.

4. The interplay between microstructure and machining parameters can be expected to have a unique effect on surface texture formation because of its unique effect on the micro-cracking activities during machining. Further research in this direction is urgently needed.

REFERENCES

Box, G., et al. 1978. *Statistics for Experimenters: An Introduction to Design, Data Analysis, and Model Building.* Wiley, New York.

Evan, A. G., and Marshall, D. B. 1980. Wear mechanisms in ceramics. In Rigney, D. A., ed., *Fundamentals of Friction and Wear.* Amer. Soc. of Metals, Metals Park, OH, pp. 439–452.

Goldstein, J., Newbury, D., and Echlin, P. 1992. *Scanning Electron Microscopy and X-Ray Microanalysis,* 2nd ed. Plenum Press, New York.

Groenou, B., Maan, N., and Veldkamp, J. 1979. In Hockey, B., and Rice, R., *The Science of Ceramic Machining and Surface Finish.* National Bureau of Standards, SP-562, p. 43.

Inasaki, I. 1987. Grinding of hard and brittle materials. *Ann. CIRP* **36**.

Jahanmir, S., Ives, L., and Ruff, A. 1992. Ceramic machining: Assessment of current practice and research needs in the United States. National Institute of Standards and Technology, Report SP-834.

Konig, W., et al. 1991. Machining of New Materials. *Proc. Adv. Mat.* 1:11–26.

Lawn, B. 1994, *Fracture of Brittle Solids.* 2nd ed. Cambridge University Press, Cambridge.

Mazurkiewicz, M. 1991. Understanding abrasive waterjet performance. *Mach. Tech.* **2**:1–3.

Nakagawa, T., Suzuki, K., and Uematsu, T. 1985. Three dimensional creep feed grinding of ceramics by machining center. In Subramanian, K., and Komanduri, R., eds., *Report in Machining of Ceramic Materials and Components.* ASME, New York, pp. 1–7.

Slotwinski, J., et al. 1993. Ultrasonic measurement of surface and subsurface structure in ceramics. *Proc. Int. Conf. Machining of Advanced Materials,* Maryland, pp. 117–124.

Tower Oil & Technology Co. 1994. *LS-A-14H Specification.* Chicago, IL.

Zhang, G., Satish, K., and Ko, W. 1994. The mechanics of material removal mechanisms in the machining of ceramics. *ASME Winter Annual Meeting Proc.* pp. 121–135.

SUGGESTED READINGS

Rekow, D., Zhang, G., and Thompson, V. 1993. Machining ceramic material for dental restorations. *Proc. Int. Conf. Machining of Advanced Materials*, The National Institute of Standards and Technology, Maryland, pp. 425–435.

Stout, K. J., ed. 1994. *Three Dimensional Surface Topography: Measurement, Interpretation, and Applications.* Penton Press: London.

Zhang, G., Ko, W., and Ng, S. 1995. Submerged precision machining of ceramic material. *ASME Conf. Advanced Material Processing*, Los Angeles.

Zhang, G., Ng, S., and Le, D. T. 1996. *Characterization of Surface Texture Formed during Machining of Ceramics.* Technical papers of the North American Manufacturing Research Institution of the Society of Manufacturing Engineers, pp. 57–62.

Zhang, G., Ng, S., and Le, D. T. 1995. Characterization of the surface cracking formed during the machining of ceramic material. *ASME Winter Annual Meeting Proc.*, vol. 1, pp. 415–429.

NEW METHODS IN NONLINEAR DYNAMICS AND CONTROL

9

NONLINEAR DYNAMICS
WITH IMPACTS
AND FRICTION

S. W. SHAW and B. F. FEENY

9.1 INTRODUCTION

Manufacturing processes often involve cutting or other sliding between contact surfaces with the possibility of separation. Frictional and impact forces can play a large role in these dynamical processes and can induce vibrations, leading to surface variability and noise. This chapter introduces the reader to some recent developments in the understanding and analysis of mechanical systems whose response includes the effects of impacts and/or dry friction. The reason for grouping these together is that they represent severe, nonsmooth nonlinearities that can cause a system to behave very differently from a smooth system. In addition these types of nonlinearities are quite prevalent in mechanical systems. Furthermore, because of their special character, systems subjected to such forces are amenable to special types of analysis. Our aim here is to introduce, in a gentle manner, some of the dynamics encountered in these systems, and some of the techniques used to study them.

The breadth of application for these problems is immense, and we will point to survey papers where the reader can further explore related topics as desired. For present purposes we will focus on relatively simple models, general features of their dynamic response, and methods of analysis. We begin with impact problems and then turn our attention to friction problems. Our focus on impact problems is on dynamic bifurcations and geometrical interpretations of the various responses encountered. For friction problems the coverage is more

Dynamics and Chaos in Manufacturing Processes, Edited by Francis C. Moon.
ISBN 0-471-15293-5 © 1998 John Wiley & Sons, Inc.

focused on chaos and techniques for measuring and describing it. In both cases the special nature of the discontinuous nonlinearity is exploited. In the closing, we summarize and also give our view regarding the value of using a dynamical systems approach for studying this class of problems.

9.2 DYNAMICS WITH IMPACTS

9.2.1 Background

Impact dynamics is a very rich and subtle field in which there still exist debates about fundamental modeling issues for impact events between rigid bodies; see, for example, Stronge (1990) and Brach (1991). However, even the simplest impact models—one-dimensional motion with simple, instantaneous velocity reflection—can exhibit extremely rich dynamical behavior. It is on this aspect of impact dynamics that we concentrate here, as it will introduce the reader to the range of possibilities for steady state behavior, as might be observed during a manufacturing process.

Simple impact systems have received much attention in recent years, especially on the complexity of response. Among these are the works of Peterka and Vacik (1992), Thompson, Bishop, and their coworkers (Bishop, 1994; Thompson and Ghaffari, 1982; Thompson et al., 1994), Shaw and coworkers (Shaw and Holmes, 1983a, b; Shaw, 1985; Moon and Shaw, 1983), Budd and Dux (1994), Nordmark (1992), Whiston (1992), and Pfeiffer and Kunert (1990). There is an entire range of studies included in this list, from detailed simulations to sophisticated bifurcation analysis of grazing motions and experiments. These show that the dynamic response of even very simple models can be extremely rich. However, the details may not be as important in applications as simply knowing the range of possibilities and having some knowledge about how these motions are affected by system parameters and characteristics. To this end we will present some typical results for the dynamics of a deceptively simple single-degree-of-freedom system.

9.2.2 A Sample Problem

Consider the simple mechanical system depicted in Figure 9.1. This represents the dynamics of a vibration system that is captured by a single mode, whose response is constrained in amplitude by a rigid barrier. For example, this could represent a crude model for a flexible-machine-tool/rigid-workpiece interaction. This system can represent devices with clearance types of nonlinearities, as shown in the picture (with a positive gap) or those pressed together by a preload, obtained by simply incorporating a "negative" gap, with $\Delta < 0$. The preload case is most relevant to the problems of interest in this volume. We provide a brief overview of the dynamic behavior, pointing out some special types of responses. Some techniques for analyzing these systems will then be

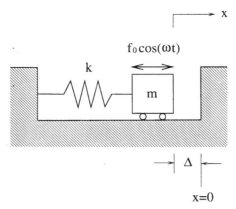

FIGURE 9.1. Mechanics model for a forced oscillator with clearance or preload. The barrier is rigid and placed at $x = 0$. The gap prameter Δ is positive for clearance and negative for preload.

outlined. The reader will be directed to specific references for more detailed studies.

The equation of motion for this system, when rescaled in terms of time and force parameters, can be written as

$$\ddot{x} + 2\zeta\dot{x} + x = -\mu + \cos(\Omega t + \phi), \qquad x < 0,$$

$$\dot{x} \rightarrow -r\dot{x} \qquad \text{at } x = 0,$$

where x represents the displacement of the mass, ζ is the damping ratio, $\cos(\Omega t + \phi)$ represents a harmonic excitation with phase ϕ, μ is the ratio of the preload force in the spring to the amplitude of the external force, and r is a simple coefficient of restitution for the impact. (Note that $\mu < 0$ is the preload case.)

The undamped, unforced dynamics of this system have periods of oscillation that range from zero for very small initial energies to π for very large initial energies (in which the preload is barely noticeable). The transient motion of this system, when it is unforced and dissipative effects are included, consists of a simple "bouncing" which, due to energy loss, decays to zero. This process, shown in the $(x, \dot{x})$ phase plane of Figure 9.2, is known as settle-out, or *chatter*, and it involves an infinite number of impacts which are completed in a finite duration for $\mu < 0$. Without further disturbance, the system will happily remain in its rest configuration.

In contrast, the dynamic response of this system when it is subjected to periodic excitation, as it might be in an environment with rotating machinery, can be extremely complicated. A few of the more important things to note about the overall system response are the following: (1) the steady state

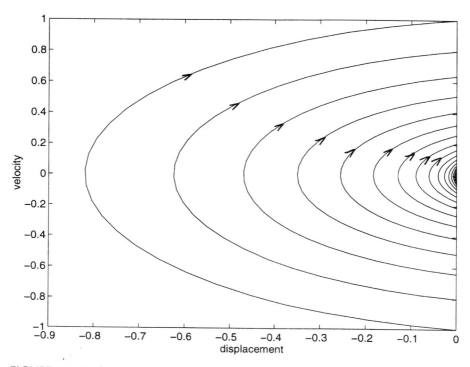

FIGURE 9.2. Unforced transient dynamics in the $(x, \dot{x})$ phase plane for a preload case: $\mu = -0.2$, $\zeta = 0$, $r = 0.8$. The system chatters to rest in finite time, undergoing an infinite number of impacts. Note that each impact at $x = 0$ involves a sudden reversal of velocity, $\dot{x} \rightarrow -r\dot{x}$.

response to a given excitation may not be unique—it may depend on initial conditions (this is a rather general feature of nonlinear systems); (2) the steady state response can be nontrivial even when the excitation levels do not overcome the preload—this is the case when the rest configuration *coexists* with one or more dynamic response(s); (3) a steady state dynamic response may not be of the same period as the excitation, and it may in fact be chaotic.

If one thinks of this system from a linear perspective, one would conclude that for $\mu < -1$ no *steady state* dynamic motion is possible, since this is the case in which the preload force is larger than the peak amplitude of the external excitation. For $\mu < -1$ the mass, if started at the rest position, will remain there for all time. However, if the system is disturbed, it may lock into a dynamic, bouncing motion. The outcome of a given disturbance—that is, whether the system settles back to rest or goes to a dynamic steady state—can depend on the initial conditions provided by that disturbance. An example of such a stable, dynamic steady-state response is shown in the phase plane in Figure 9.3. Figure 9.3 shows a relatively large amplitude bouncing motion that

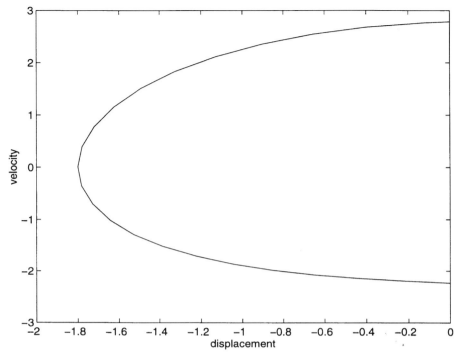

FIGURE 9.3. At least two possible stable steady state solutions coexist at the following set of parameter values: $\mu = -1.1$, $\zeta = 0$, $r = 0.8$, $\Omega = 2.8$. In addition to the rest configuration that exists since $\mu < -1$, the simple periodic bouncing motion shown here also exists. Initial conditions dictate which one will be observed.

repeats after each impact. It should be noted that the rest configuration of this system is also stable steady state response for this set of parameters.

Similarly, for the case of clearance, $\mu > 0$, a dynamically stable, impacting steady state can coexist with the purely linear steady state response. This is demonstrated in Figure 9.4. Again, if one thinks in purely linear terms, this possibility is overlooked. (The recent paper by Thompson et al., 1994, provides a nice summary of these motions for clearance systems.) Another possibility is that these systems can exhibit sustained chaotic behavior. An example of this is given in Figure 9.5, where the preload in this case is zero.

For design and control purposes, it is desirable to be able to determine the sources of these types of responses and to be able to predict when they occur in terms of system and excitation parameters. Even for this simple system, a complete answer to such central questions is far from possible. However, the geometrical tools of dynamical systems have proved to be very helpful for providing partial answers, and they can offer valuable insights into why these problems are difficult.

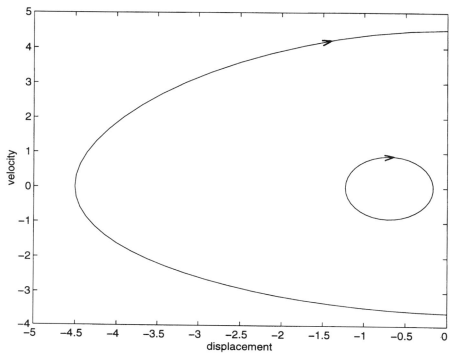

FIGURE 9.4. A clearance case in which the linear steady state coexists with a simple bouncing response; $\mu = 0.7$, $\zeta = 0$, $r = 0.8$, $\Omega = 1.7$.

9.2.3 Methods of Analysis

To attack this problem, or similar problems with abrupt changes in character, it is natural to exploit the nature of the phase space. The usual states of this system are the displacement and velocity of the mass $(x, \dot{x})$. However, since the system has an explicit, periodic time dependence in the excitation term, it is useful to "suspend" the system by replacing t by a variable θ and adding the trivial differential equation $\dot{\theta} = 1$ to the dynamical system. The state space for this augmented system is the plane in which $(x, \dot{x})$ exist *plus* the phase associated with θ. (Note that since the θ dependence of the equation of motion is periodic with period $T = 2\pi/\Omega$, only its phase is important.) One can cut this three-dimensional phase space with a two-dimensional surface, called a *surface of section*, or *Poincaré section*, and consider the points on that surface that arise from intersections with solutions of the differential equation as time evolves. The mathematical rule that describes the behavior of these points is referred to as a *Poincaré map*. Such maps are powerful tools for describing qualitative features of low-order dynamic systems — especially complicated time-dependent behavior such as chaos. However, they are rarely available in

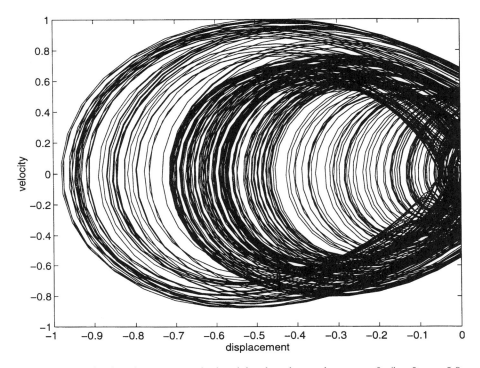

FIGURE 9.5. A chaotic response depicted in the phase plane; $\mu = 0$, $\zeta = 0$, $r = 0.8$, $\Omega = 2.8$.

analytical form and therefore have limited usefulness as a predictive tool. An exception is the class of systems under consideration here, where one can often get some useful information about the system by considering its Poincaré map.

For systems described by piecewise-linear or other nonsmooth characteristics, a natural place to construct the surface of section is along the surface that separates the phase space into its various parts. By making use of the solutions of the differential equations that are valid in each domain and employing appropriate matching conditions, information regarding certain classes of responses can be obtained in an analytical manner. For example, the existence and stability of many periodic solutions can be obtained. In addition efficient simulations can be tailored to this class of problems.

For the sample problem under consideration, the abrupt change at $x = 0$ offers the natural cut in the phase space. The Poincaré section is taken to be the surface given by points in the phase space corresponding to the mass just coming into contact with the barrier, that is, at $x = 0$, $\dot{x} > 0$. In this surface the states $(\dot{x}, \theta)$ — the velocity and forcing phase just prior to impact — are measured at each impact during a motion. This surface is shown in Figure 9.6. Periodic motions such as those depicted in Figures 9.3 and 9.4 are represented

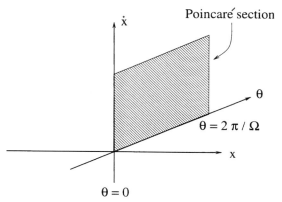

FIGURE 9.6. The extended phase space, $(x, \dot{x}, \theta)$ and the Poincaré section (shaded).

by a single point in this surface (recall that we record only the phase of the time by θ), since they repeat after each impact. Such a point is called a *fixed point* of the Poincaré map.

This fixed point, and its stability type, can be found in closed form for this system as follows: First, we express the solution for a motion that leaves an impact at time $t = t_0$ with velocity $y_0 (<0)$ as $x(t; t_0, y_0)$. The next impact occurs at a time t_1 with velocity y_1, and this event can be expressed by

$$x(t_1; t_0, y_0) = 0, \quad \dot{x}(t_1; t_0, y_0) = y_1.$$

If one could use these equations to solve for (t_1, y_1) as a function of (t_0, y_0) and the system parameters, these functions, combined with the simple impact rule, would represent the Poincaré map for this system (for the given surface of section). Unfortunately, the solution involves transcendental equations and is not available. However, periodic motions which repeat after each impact can be captured analytically. For the motion to repeat itself after each impact, it must hold that

$$t_1 = t_0 + \frac{2\pi n}{\Omega} \quad \text{and} \quad y_1 = -\frac{y_0}{r}.$$

In this manner the velocity after impact at $t = t_1$ is again y_0, and the forcing phase is the same as that at the previous impact, although n periods of the forcing have transpired. (Such a motion for $n > 1$ is called a *subharmonic of order n*.) Since we have a closed-form solution for $x(t; t_0, y_0)$, this can be carried out analytically. For the present case, since we have an arbitrary phase in the excitation, we can take $t_0 = 0$ and solve for the y_0 and ϕ values corresponding to the periodic response. The details for such a procedure can be found in many of the papers on impact oscillators listed below, and even as far back as the work of Den Hartog (1931) on friction problems. It yields a quadratic equation for y_0 that is uncoupled from the equation for ϕ, and solving these provides the fixed point location in the Poincaré section. This fixed point contains expressions for the impact velocity and forcing phase at impact for single-impact periodic motions. Impact-velocity versus frequency response curves for different levels of μ are shown in Figure 9.7. These are simply solution branches that match the periodicity conditions. An important question is that of the dynamic stability of these responses.

By tracking the dynamics of a small perturbation of the velocity and phase near a fixed point, and studying its linear, discrete-time dynamics, the stability type of a fixed point can be determined. By examining changes in stability, bifurcations are found in which the qualitative nature of the response changes as a parameter is varied through a critical, or bifurcation, parameter value. These bifurcations include the usual ones encountered in smooth systems—such as saddle-node or period-doubling—and also the *grazing bifurcations*, which are unique to nonsmooth systems (more on this below). It is not appropriate to go into the details of these bifurcations here, but we can demonstrate some of them for this sample problem. In Figure 9.7 saddle-node bifurcations are evident in which two periodic-response branches merge and annihilate one another as the input frequency is varied. (This is the classical turning-point, or vertical tangency, bifurcation.) Figure 9.8 demonstrates a typical period-doubling bifurcation. Some bifurcations, such as the period-

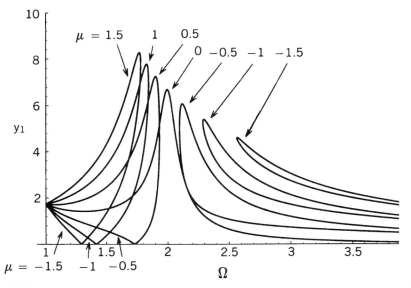

FIGURE 9.7. Impact velocity versus forcing frequency for various levels of μ for $r = 0.8$. Stability type is not indicated, nor have nonphysical solutions been omitted.

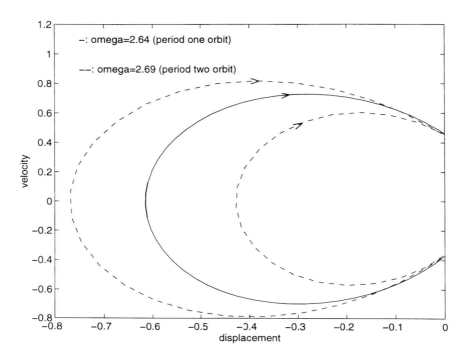

FIGURE 9.8. The chaotic motion of Figure 9.5 as viewed in the Poincaré section; $\mu = \theta$, $\zeta = 0$, $r = 0.8$, $\Omega = 2.8$.

doubling, are rather benign in terms of the gross behavior of the system — that is, they do not cause a drastic change in the amplitudes, stress levels, and so on. However, the saddle-node and grazing bifurcations can be quite dramatic.

When studying these periodic responses, it is important to note that the mathematical conditions for the existence of a fixed point can, and often do, yield solutions that are nonphysical. This is due to the fact that they do not ensure that the solution remains in $x < 0$ during the entire periodic cycle. One must be very careful about keeping these spurious solutions in mind during the analysis, although they cannot be omitted in any simple manner. The transition from motions that are physical to those that penetrate the barrier involve *grazing bifurcations*, which involve motions in which the mass strikes the wall with zero velocity. These are extremely important in the overall dynamics of impact oscillators and have received much recent attention (Nordmark, 1991 and Thompson et al., 1994).

The Poincaré section can be also used to depict complex motions. For example, the response shown as the untidy collection of trajectories in Figure 9.5, appears in the Poincaré section as the collection of dots shown in Figure 9.9. This structure of dots is referred to as a *strange attractor*, and it has a fractal structure (Moon, 1992). Furthermore it can be observed that this structure lies on the *unstable manifold* of a saddle-type fixed point of the

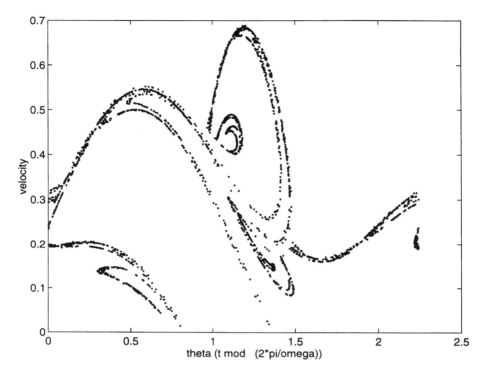

FIGURE 9.9. A period-doubling bifurcation for $\mu = 0$, $\zeta = \theta$, $r = 0.8$. There are two separate steady states shown; the one for $\Omega = 2.64$ repeats after each impact (solid line), while the one for $\Omega = 2.69$ repeats after alternate impacts (dashed lines).

Poincaré map, which can be found by the analysis outlined above (see Shaw and Holmes, 1983a).

As parameters are varied, one may encounter various sequences of bifurcations, including the usual period-doubling route to chaos, as well as grazing transitions that lead to sudden, drastic changes in the response as parameters are varied. These result in very complicated frequency responses for this class of systems.

In closing this section it must be pointed out that we have considered only the very simplest impact model, yet it yields very rich behavior. Therefore it can be expected that more complex models will be at least as equally rich.

9.3 DYNAMICS WITH FRICTION

In this section we present some of the nonlinear phenomena associated with friction, and some of the methods and thought processes that go into the analysis. We start with some comments on friction modeling. We then focus on the nonlinear dynamics associated with stick-slip motion.

9.3.1 Friction Modeling

The details of the friction process are not very well understood. Hence there is no universal friction law that is applicable in every study of friction. Instead, models vary according to the application and the phenomena at hand. In this section we briefly review some of the friction models currently in use.

Basic elements of friction modeling go back to Leonardo da Vinci, who stated that "friction produces double the amount of effort if the weight is doubled" and established essentially that friction was independent of the contact area. These comments are consistent with our commonly used notion that the friction is proportional to the normal load and that the proportionality constant is the coefficient of friction. These ideas were forgotten and then reformulated by Amontons in 1699, and later by Coulomb in 1781.

The use of a constant coefficient of friction during sliding is commonly called the *Coulomb friction law*. Sometimes included with Coulomb's law is the effect of a static coefficient of friction. Hence the friction law is that $f = \mu(v)N$, where N is the normal load, v is the relative velocity, and the coefficient of friction obeys $\mu(v) = \mu_k \operatorname{sign}(v)$, $v \neq 0$, and $|\mu(0)| \leqslant \mu_s$. This effect can model the infinitely many equilibria and the stick-slip motion encountered in oscillators with friction. The difference between static and kinetic friction can also model the capability of self-excited stick-slip oscillations. However, this simple friction law causes mathematical difficulties because it is discontinuous and multivalued at $v = 0$.

Stribeck friction also includes static friction, and a multivalued discontinuity at $v = 0$. This law incorporates a continuously decreasing friction magnitude as the magnitude of the velocity increases. If $|v|$ is larger than some value, the friction may be allowed to increase, although this is not necessary. This friction law also models the infinite locus of equilibria in oscillators, as well as self-excited oscillations. It predicts that in a belt-driven system the static equilibrium may be unstable, in which case smooth oscillations grow until they reach stick-slip limit cycles.

While the above simple friction laws predict instabilities based on the friction characteristic alone, it has been observed that such instabilities can also depend on stiffnesses, normal loads, and angles of contact. An example is the chalk on the blackboard. By common experience we know that the angle at which the chalk is held to the blackboard and the normal force are critical in determining whether it will chatter or screech. (Also try rubbing your fingertip on the tabletop.)

Another mechanism for generating self-excited vibrations is the elastic interactions of the contact structures (D'Souza and Dweib, 1990; Oden and Martins, 1985; Tworzydlo et al., 1992). Researchers using these models have demonstrated the dependence of instability on stiffness, normal load, and angle of contact. Such dependencies have been observed by many researchers (see Ibrahim, 1994, for a review).

Hysteretic friction behaviors have been observed in sliding rocks and metals

(Dieterich, 1979; Ruina, 1985; Dupont and Kasturi, 1995). The basic feature is that when the sliding velocity increases suddenly, the friction force also quickly increases, and then settles to a lower value. An additional state variable has been used to empirically model this feature.

9.3.2 Existence and Uniqueness

It was mentioned above that friction models are typically discontinuous at a relative velocity of zero. This leads to a discontinuity in the differential equations that describe the motion a frictional system. Thus the basic theorems of existence and uniqueness become issues. In a simple mass-spring system, the broad range of the static friction force at zero velocity, at the discontinuity, is necessary for the existence of solutions. This multivaluedness in the friction model must be large enough to "join" the discontinuous pieces. Indeed this is consistent with Filippov's (1964) construction of solutions for discontinuous systems, in which the vector field at the discontinuity is assigned according to an algorithm. In other systems uniqueness of solutions can also be an issue when dealing with friction (Painlevé, 1895; Lötstedt, 1981; Dupont, 1992).

9.3.3 Stick-Slip Oscillations

Throughout the rest of this chapter, we will focus on the nonlinear dynamics associated with stick-slip motions. Stick-slip was observed as early as in the 1860s by Helmholtz (1954) in his studies of bowed strings, in which he noted that "During the greater part of each oscillation the string here clings to the bow, and is carried on it; then it suddenly detaches itself and rebounds, whereupon it is seized by other points in the bow and carried again forward."

We will study stick-slip through models using the simple friction laws. However, analogous phenomena should carry over to more complicated friction models, as long as they can accommodate sticking behavior.

Example

A standard example is a forced mass-spring system with Coulomb damping (Figure 9.10). The nondimensionalized equation of motion is

$$\ddot{x} + 2\zeta\dot{x} + x + n(x) f(\dot{x}) = a \cos \Omega t, \tag{1}$$

where x is the displacement, ζ is the damping ratio, and $a \cos \Omega t$ represents a harmonic excitation. $f(\dot{x})$ represents the coefficient of friction, and is given by $f(\dot{x}) = \mu \, \text{sign}(\dot{x})$ for $\dot{x} \neq 0$, and $-1 \leqslant f(\dot{x}) \leqslant 1$ for $\dot{x} = 0$. For a constant normal load, $n(x) = 1$. Periodic responses were analyzed by Den Hartog (1931). Shaw (1986) extended this work with a modern stability analysis, and noticed that the stick-slip dynamics reduced to a one-dimensional map.

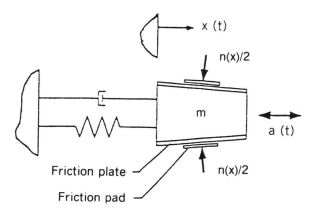

FIGURE 9.10. Mechanics model for a forced oscillator with dry friction. The friction plates are fixed to the mass m and slide relative to the friction pads, which are fixed in x. The friction surfaces are not parallel in the direction of displacement x. Thus the elastically loaded normal forces vary with displacement.

Allowing the normal load to vary linearly with displacement, we have

$$n(x) = 1 + kx, \qquad x > -\frac{1}{k},$$

$$n(x) = 0, \qquad x < -\frac{1}{k}. \tag{3}$$

To prevent the existence of a negative normal load (and negative friction), the model allows for a loss of contact. This oscillator can undergo stick-slip chaos on a "branched manifold," which is a folded sheet in the phase space upon which the solutions, or trajectories, reside. Figure 9.11 shows a phase portrait of this system in cylindrical coordinates. The radial axis represents the displacement, the longitudinal axis corresponds to the velocity, and the circumferential axis is time. Trajectories evolve clockwise. The plane $\dot{x} = 0$ includes motions that are momentarily stuck. Trajectories near the outer portion of the chaotic attractor wind around the time axis and move slightly inward when the mass slips. Trajectories on the inner portion of the image slip greatly as they navigate the fold in the attractor. Outer trajectories correspond to large displacements, and therefore larger friction forces and less slip. Inner trajectories correspond to smaller displacements, and therefore lighter friction and greater slip.

A Poincaré section, in this case a cross section of the phase trajectories at a particular phase of excitation, reveals a 1-D structure embedded in 2-D space (Figure 9.12a). By assigning a coordinate axis s along this 1-D structure, each point is associated with a value of s. We can form a return map by plotting s_{n+1} versus s_n. This exposes an underlying deterministic function that rules the

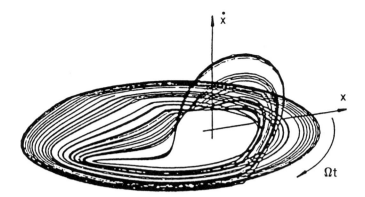

FIGURE 9.11. Three-dimensional phase portrait of Equations (1) and (3) with $a = 1.9$, $\Omega = 1.25$, $k = 1.5$, and $\zeta = 0$. In cylindrical coordinates, x is the radial axis, $\dot{x}$ is the longitudinal axis, and t is the circumferential coordinate.

dynamics via a 1-D map (Figure 9.12b). Thus there is a single-humped 1-D map underlying the dynamics (Feeny and Moon, 1994; Feeny, 1992). Period doubling was the observed route to chaos as parameters were varied.

An experimental rendition of this model consisted of a mass attached to an elastic beam. The mass had titanium plates on both sides, providing friction surfaces. Spring-loaded titanium pads rested against the titanium plates. The plates had an angle between them so that the normal load would vary with displacement. Details on the experiment can be found in Feeny and Moon (1994). As with the model, the Poincaré section looks like a (somewhat noisy) 1-D image bent in 2-D space (Figure 9.13a). The crisp part of the image corresponds to sticking motion. This indicates that there is heavy dissipation during a stick. A physical interpretation is that the excitation energy is

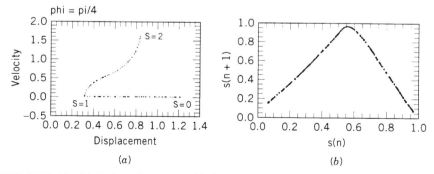

FIGURE 9.12. (a) Poincaré section. (b) A return map on the coordinate s shows the underlying one-dimensional map.

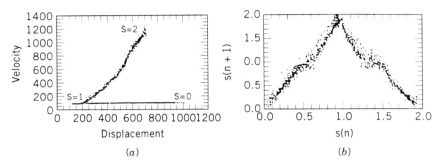

FIGURE 9.13. (a) Experimental Poincaré section. The crisp part corresponds to sticking, the fuzzy part to slipping. (b) A return map on the coordinate s shows the underlying one-dimensional map.

absorbed while the oscillator is stuck. A return map (Figure 9.13b) exposes the underlying 1-D map. The level regions at $s = 0.5$ and $s = 1.5$, and perhaps the sharp maximum, are artifacts of the manner in which the coordinate s was assigned through the bent part of the Poincaré image. As with the model, period doubling was observed.

The model of a belt-driven forced oscillator also underges stick-slip chaos (Popp and Stelter, 1990; Popp, 1992; Popp et al., 1996). It has an underlying 1-D map that resembles a circle map. As with circle maps, the oscillator exhibits intermittency and period doubling as routes to chaos.

The upshot of these observations is that stick-slip motions for single-degree-of-freedom oscillators is a class of motion in which the underlying dynamics can be captured nicely by a one-dimensional map. The remainder of this chapter focuses on oscillators modeled by Equations (1) and (3).

9.3.4 The Geometry of Stick-Slip

It is easy to imagine how sticking occurs. If we have a mass mounted to the wall by a spring, we can place the mass in many positions, and it will remain stuck, as long as the spring force is less than the friction force. However, if we give the mass enough displacement, such that the spring force exceeds the static friction, the mass will slide. If there is an externally applied time-varying force, the mass will be stuck until the force exceeds the static friction so that the mass slips.

Here we discuss mathematically how the multivalued discontinuity can model stick-slip and lead to some interesting geometric properties of motion.

The discontinuity poses some limitations in traditional analyses and thus challenges the dynamical systems community. We cannot take derivatives of the discontinuous model. Thus linearizations about the discontinuity, and variational equations if they involve the discontinuity, are of no use in analysis. Any dynamical systems theorems that call for a "smooth vector field," a "smooth manifold," or a "hyperbolic fixed point" must be questioned in stick-slip (or impact) systems.

Stick-slip can be explained as a result of the multivalued discontinuity in the friction function $f(\dot{x})$. From the differential equations of motion, such as Equation (1), we can uncover *sticking regions*. This can be done by evaluating the equation of motion for fixed points (Shaw, 1986; Feeny and Moon, 1994) and taking advantage of the continuous range of the friction force at zero velocity, $-1 < f(0) < 1$. Because of the inequality there is a range of values at which equilibrium can hold, giving rise to a sticking region. (In belt-driven oscillators, instead of seeking fixed points, we seek a solution at which $\dot{x} = v$, where v is the velocity of the belt. This becomes a constraint that holds only while the friction force is balanced.)

Another way to analyze the sticking region is by evaluating the direction of the vector field on both sides of the surface of discontinuity D, which is defined by $\dot{x} = 0$. The sticking region exists where both piecewise vector fields adjacent to either side of D, and with the friction at its static value, point *into* D. Where the sense of both piecewise vector fields agree with respect to D, solution trajectories simply pass through D. Where both piecewise vector fields point away from D, we would have an unstable sticking region which could never be maintained.

Let us imagine what happens as a 3-D "blob" of solution trajectories approaches the sticking region. (This is a typical approach in dynamical systems analysis—to consider the qualitative behavior of collections of solutions.) We know that this blob is unable to pass through D, so it gets completely compressed onto D, since all information about velocity is compressed onto zero velocity. What started as a 3-D set has been compressed into a 2-D set. In the 2-D Poincaré section taken at a phase of the forcing period, this 2-D set is seen as a 1-D line. This illustrates the source of the underlying 1-D map.

A consequence of this singular compression of the 3-D set to a 2-D set is that the process that sends motions through a sticking region is noninvertible. Another consequence is that if the attractor consists of sticking motions, then it is possible to reach the attractor in finite time, during one sudden collapse in the sticking region. The 1-D map which describes the attractor is then an exact representation. Extrapolating the map back into the full phase space, we obtain an attractor represented by an object of dimension two or less in the 3-D phase space. (In the numerical example, the correlation dimension was estimated as 1.93 ± 0.04.)

These effects of the discontinuity can also have detrimental effects on control (for a comprehensive review, see Armstrong-Hélouvry et al., 1994).

9.3.5   The Use of One-Dimensional Maps: Symbol Dynamics

Since one-degree-of-freedom stick-slip oscillators can be represented by one-dimensional maps, the Eldorado of mathematical results on 1-D maps becomes applicable. Since this chapter focuses on a particular oscillator associated with a single-humped map, we provide some background on this map (see Devaney, 1987, or Guckenheimer and Holmes, 1983, for more details).

We consider maps on the unit interval $I = [0, 1]$ of the form $x_{n+1} = \lambda g(x_n)$, where $g(0) = g(1) = 0$, $g(x)$ has a smooth maximum at $x = x_0$, $g(x) > 0$ for $0 < x < 1$, and λ is a bifurcation parameter such that $\lambda \leqslant \lambda_{\text{crit}} = 1/g(x_0)$.

The dynamics of such 1-D "hump" maps can be characterized by *symbol sequences*. A symbol sequence for a motion can be defined by whether the ith iterate is to the right of x_0, in which case we take the symbol R, or to the left of x_0, in which case we take the symbol L. There is a direct relationship between a real number $x \in I$ and the symbol sequence generated by using it as an initial condition. For example, for $\lambda g(x) = 4x(1 - x)$, there is a one-to-one correspondence between $x \in I$ and symbol sequences. The attractor is the entire interval I, and any sequence is a possible sequence in this map. These sequences are a powerful tool in proving results about these maps.

Stick-slip provides a natural basis for defining a symbol sequence. If the Poincaré section is chosen correctly, then the symbol dynamics of the 1-D map correspond identically to the symbols corresponding to stick and slip (Feeny and Moon, 1993). Thus the corresponding symbol sequence would be S for *sticking*, and N for *not sticking*.

It is possible to perform some data analysis directly on the symbol sequence. Singh and Joseph (1989) developed a method of easily estimating the order of magnitude of the maximum Lyapunov exponent. The Lyapunov exponent measures the local divergence of neighboring trajectories. (See the chapters by Abarbanel and Moon for more details.) A positive Lyapunov exponent indicates sensitive dependence on initial conditions, or chaos, if the motion remains bounded. From a binary autocorrelation function for an N-sequence of 1's and -1's with a mean of zero, it is possible to estimate the macrscopic Lyapunov exponent. (The exponent is dubbed "macroscopic," since the resolution of a binary partition prevents us from examining expansions and contractions at an infinitesimal scale.) This idea has been successfully applied to stick-slip symbol-sequence data from the experimental and mathematical friction oscillator described above (Feeny and Moon, 1989).

Under certain assumptions there is a "universal sequence" of bifurcations of periodic orbits as λ increases in the range $0 \leqslant \lambda \leqslant \lambda_{\text{crit}}$. Not only will all such maps produce the same sequence of periodicities, but each periodic cycle in the sequence will always have a certain symbol sequence as well. It is thus possible to investigate the universality of the underlying map by closely examining its bifurcation sequence and the associated symbol sequence. As a result one can obtain information as to whether the system matches a certain class of assumptions. The bifurcation sequence of the Coulomb oscillator matches the universal sequence, although some expected segments have not been observed (Feeny and Moon, 1993). This is somewhat of an inconclusive result, since it is always possible that bifurcation events are taking place on parameter windows that are too small for a practical search. However, in this case it is possible to show that the 1-D map intrinsic to the Coulomb oscillator does not satisfy all of the basic hypotheses necessary for universality (Feeny and Moon, 1993; Feeny, 1996).

9.3.6 On the Reconstruction of Stick-Slip Data

A handy tool for experimentalists in nonlinear dynamics and chaos is the *phase-space reconstruction* in which information on the entire system is extracted from a single observable. The intuitive idea is that all states are coupled together and therefore influence each other. Thus the observed time series data should contain information about all of the active states of a system. A typical way to recover this information is to map the observed one-dimensional time history into an E-dimensional space. When it is possible to do this in a continuous, invertible manner, this is called an *embedding*. In the embedding space the data will approximate a limit set that is in some way equivalent to the limit set in the original phase space. Details on how to do this properly are reviewed by Abarbanel (1995).

Given a reconstructed phase space, a researcher can then address the dimensionality of the dynamical system, which is helpful in the initial modeling stages. This has been successfully applied in cutting processes (Moon and Abarbanel, 1996).

Typically phase-space reconstructions rely on the *method of delays*. Given a sampled time history y_n, E-tuples of the data spaced by a delay k are collected into E-dimensional vectors $(y_n, y_{n+k}, \ldots, y_{n+Ek})$. These vectors are then plotted sequentially to form orbits in this "constructed" phase space. Takens' embedding theorem (Takens, 1981) says that if basic hypotheses are met, then the method of delays performs an embedding. One of these hypotheses is that the observable satisfy a smoothness criteria: $y \in C^r$, where $r \geqslant 2$. If we choose $y = x$, then the discontinuous acceleration (due to discontinuous forces) implies that $y \notin C^2$. Should we choose the observable $y = \dot{x}$, then $y \notin C^1$. Hence the frictional discontinuity in the vector field gives our system the opportunity to disobey this theorem. And it does.

Problems occur when orbits pass through the sticking region. During a stick interval that starts at x_i, we have m samples of constant $x_j = x_s$. In a 3-D reconstruction we build vectors that include

$$(x_i, x_{i+k}, x_{i+2k}), \ldots, (x_{i+m-2k}, x_{i+m-k}, x_{i+m}),$$

all of which are equal to (x_s, x_s, x_s). Thus, when we plot all of these vectors into the reconstructed 3-D space, they all pile up on the same point on the identity line. This is shown for numerical data in Figure 9.14. The figure shows that the sticking region in the 3-D numerical solution of Equation (1) collapses during a reconstruction from the x variable. Thus the method of delays gives rise to a singularity in mapping the displacement history (x_j) to the 3-D reconstructed phase space (Feeny and Liang, 1995).

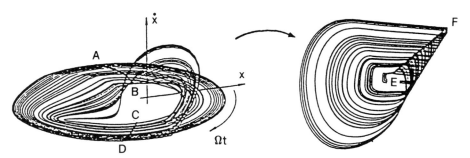

FIGURE 9.14. Reconstructed limit sets may not be similar to limit sets in the original phase space. Points in the sticking region between AB and CD effectively get crushed onto the line EP, which coincides with the identity line in the pseudo phase space.

This affects reconstruction-based computations that characterize strange attractors, such as Lyapunov exponents and dimension calculations. While reconstructions and dimensionality studies have been successful in cutting processes, care should be taken if stick-slip is involved in the dynamics.

9.3.7 Summary

Stick-slip and impact processes correspond to discontinuities in mathematical models. This is a special feature, and it places such oscillators into a special class of mechanical systems. One of the fundamental characteristics of single-degree-of-freedom oscillators is that their dynamics can be conveniently represented by mappings. For single-degree-of-freedom friction oscillators these can be one-dimensional maps, due to the dimensional collapse that takes place in the state space during a stick. The dimensional collapse has other consequences: a noninvertible flow, attractors reached in finite time, and strange attractors with dimensions of two or less. We focused on one of the oscillators, represented by a single-humped map, and discussed its relationship to some analytical and computational tools.

Finally we discussed some problems that stick-slip processes pose for the analysis of experimental data. The method of delays can fail in the phase-space reconstruction of a stick-slip process. This would then affect all analyses that are based on the reconstructed data. Such analyses might include the computation of Lyapunov exponents, fractal dimensions, rotations of unstable periodic orbits, and nonlinear prediction.

Stick-slip dynamics present very interesting challenges for the mathematical analysis of dynamical systems and for the geometric techniques in data analysis. These have a direct bearing on their application to many manufacturing problems, since friction plays a central role in the tool-workpiece interaction.

9.4 CLOSING

Here we outline some questions we expect that the reader may have raised during his or her reading, and our replies to these questions. In this way we also offer a view of the role of dynamical systems theory in application areas such as manufacturing.

- Are these behaviors generic in any way, or are they special to the models considered here? The answer to this is that while the details are very problem-dependent, the behaviors reflect generic patterns of response.
- Is the nature of the impact rule important? Again, yes for specifics, but not for qualitative features. However, if one "softens" the impact substantially such that the contact time is comparable to the free-flight time, then the nonlinear behavior is less pronounced. Also, if the impact is highly dissipative, one gets more regular behavior, although some types of chaos and bifurcations still occur (Shaw and Holmes, 1983b).
- Is the nature of the friction rule important? The capability of modeling certain phenomena depends on the choice of the friction law. Thus the chosen friction law is driven by whether the system exhibits particular behavior such as stick-slip, self-excited vibration or noise, dependence of stability on normal load and stiffness, and frictional hysteresis. However, a system with stick-slip can produce behavior anlogous to that depicted in this chapter. The methods are applicable to the lower-order systems, but the phenomena and concerns extend to higher-order systems.
- What about systems with many degrees of freedom, and/or several possible contact points? These problems are very rich and very difficult. One needs to be clever to even carry out the simulations.
- Is there any hope of obtaining a complete understanding of the dynamics of such a system, even if an accurate mathematical model is known? No, even if the system is very simple, one must combine some analysis with simulations to obtain an overall picture. This is nearly impossible if the system has many parameters and degrees of freedom.
- Given the answer above, What is the value of a dynamical systems approach? The answer to this is similar to the answer to: Why study single-degree-of-freedom linear systems? By understanding the behavior of simple systems, and knowing the possibilities of dynamic behavior, one can know at least what the possibilities are. Also the results of bifurcation theory tell us that there are generic bifurcations that occur in dynamical systems, and these can be classified. In addition, a basic understanding of chaos and its important general features, such as sensitive dependence on initial conditions or continuous spectrum, can be understood from simple models. The reader is referred to Moon (1992), Guckenheimer and Holmes (1986), or Wiggins (1990) for details on these subjects.

From an engineering point of view, the field of dynamical systems is quite mature, and it is finding useful applications in many areas of study. It has been oversold by some as a cure-all for nonlinear problems that arise in engineering and other fields. However, these geometrical/analytical approaches offer valuable insights into system dynamics and provide a powerful framework by which systems can be studied and classified. Also, for many problems the dynamical systems approach produces useful results not achievable by classical techniques, and for this reason its basic results should be included in the dynamicists' "bag of tricks." (This is especially true for the class of systems considered here, since traditional perturbation methods fall short when one encounters abrupt nonlinearities.) In this regard collaborations between nonlinear dynamicists and manufacturing engineers could prove to be a fruitful venture.

ACKNOWLEDGMENTS

The authors are grateful to Frank Moon for the invitation to contribute this chapter and to Rocky Chen and Paul Chao for their help in preparation of the manuscript.

REFERENCES

Abarbanel, H. D. I. 1995. Tools for analyzing observed chaotic data. In Guran, A., and Inman, D., eds., *Smart Structures, Nonlinear Dynamics, and Control.* Prentice Hall, Englewood Cliffs, NJ.

Armstrong-Helouvry, B., Dupont, P., and Canudas de Wit, C. 1994. A survey of models, analysis tools and compensation methods for the control of machines with friction. *Automat.* **30**:1083–1138.

Bishop, S. R. 1994. Impact oscillators. *Phil. Trans. R. Soc. London A* **347**:345–448.

Brach, R. 1991. *Mechanical Impact Dynamics.* Wiley, New York.

Budd, C., and Dux, F. 1994. Chattering and related behaviour in impact oscillators. *Phil. Trans. R. Soc. London A* **347**:365–389.

Dieterich, J. 1979. Modeling of rock friction, 1. Experimental results and constitutive equations. *J. Geophys. Res.* **84**:2161–2168.

Den Hartog, J. P. 1931. Forced vibrations with combined Coulomb and viscous friction. *Trans. ASME* **53**:107–115.

Devaney, R. L. 1987. *An Introduction to Chaotic Dynamical Systems.* Addison-Wesley, Reading, MA.

D'Souza, A., and Dweib, A. 1990. Self-excited vibrations induced by dry friction. II: Limit cycle analysis. *J. Sound Vib.* **137**:177–190.

Dupont, P. E. 1992. The effect of Coulomb friction on the existence and uniqueness of the forward dynamics problem. *Proc. IEEE Conf. Robotics and Automation,* pp. 1442–1447.

Dupont, P. E., and Kasturi, P. S. 1995. Experimental investigation of friction dynamics associated with normal load. *Proc. Design Engineering Conf.* **3A** ASME DE-Vol. 84-1, pp. 1109–1116.

Feeny, B. F. 1996. The nonlinear dynamics of oscillators with stick-slip friction. In Guran, A., Pfeiffer, F., and Popp, K., eds., *Dynamics with Friction.* World Scientific, Singapore, pp. 36–92.

Feeny, B. F., and Liang, J. W. 1995. Phase-space reconstructions of stick-slip systems. *Proc. Design Engineering Technical Conf.* **3A**, ASME DE-Vol. 84-1, pp. 1049–1060.

Feeny, B. F., and Moon, F.C. 1989. Autocorrelation on symbol dynamics for a chaotic dry-friction oscillator. *Phys. Lett. A* **141**:397–400.

Feeny, B. F., and Moon, F. C. 1993. Bifurcation sequences of a Coulomb friction oscillator. *Nonlin. Dynamics* **4**:25–37.

Feeny, B. F., and Moon, F. C. 1994. Chaos in a forced dry-friction oscillator: experiments and numerical study. *J. Sound Vib.* **170**:303–323.

Filippov, A. 1964. Differential equations with discontinuous right-hand sides. *AMS Trans.* **42** (series 2):199–231.

Guckenheimer, J., and Holmes, P. J. 1983. *Nonlinear Oscillations, Dynamical Systems, and Bifurcations of Vector Fields.* Springer-Verlag, New York.

Helmholtz, H. L. F. 1954. *On the Sensations of the Tone as a Physiological Basis for the Theory of Music,* Dover, New York.

Ibrahim, R. 1994. Friction-induced vibration, chatter, squeal, and chaos: Parts I and II. *Appl. Mech. Rev.* **47**:227–253.

Lötstedt, P. 1981. Coulomb friction in two-dimensional rigid bodies. *Z. Angew. Math. Mech.* **61**:605–615.

Moon, F. C. 1992. *Chaotic and Fractal Dynamics.* Wiley, New York.

Moon, F. C., and Abarbanel, H. 1995. Chaotic motions in normal cutting of metals. *Nature,* forthcoming.

Moon, F. C., and Shaw, S. W. 1983. Chaotic vibrations of a beam with nonlinear boundary conditions. *Int. J. Nonlin. Mech.* **18**:465–477.

Nordmark, A. 1992. Effects due to low velocity impact in mechanical oscillators. *Int. J. Bifurcation Chaos* **2**:597–605.

Oden, J. T., and Martins, J. A. C. 1985. Models and computational methods for dynamic friction phenomena. *Comp. Methods Appl. Mech. Eng.* **52**:527–634.

Painlevé, M. 1895. Sur les lois du frottement et glissement. *C. R. Acad. Sci. Paris* **121**:112–115.

Peterka, F., and Vacik, J. 1992. Transition to chaotic motion in mechanical systems with impacts. *J. Sound Vib.* **165**:95–115.

Pfeiffer, F., and Kunert, A. 1990. Rattling models from deterministic to stochastic processes. *Nonlin. Dynamics* **1**:63–74.

Popp, K. 1992. Some model problems showing stick-slip motion and chaos. In Ibrahim, R. A., and Soom, A., eds., *Friction-Induced Vibration, Chatter, Squeal and Chaos,* ASME DE-Vol. 49, pp. 1–12.

Popp, K., and Stelter, P. 1990. Nonlinear oscillations of structures induced by dry friction. In Schiehlen, W., ed., *Nonlinear Dynamics in Engineering Systems.* Springer-Verlag, Berlin.

Popp, K., Hinrichs, N., and Oestreich, M. 1996. Analysis of a self-excited friction oscillator with external excitation. In Guran, A., Pfeiffer, F., and Popp, K., eds., *Dynamics with Friction*, World Scientific, Singapore, pp. 1–35.

Ruina, A. 1985. Constitutive relations for frictional slip. In Bazant, Z., ed., *Mechanics of Geomaterials*. Wiley, New York, pp. 169–187.

Shaw, S. W. 1985. Forced vibrations of a beam with one-sided amplitude constraint—Theory and experiments. *J. Sound Vib.* **99**:199–212.

Shaw, S. W. 1986. On the dynamic response of a system with dry friction. *J. Sound Vib.* **108**:305–325.

Shaw, S. W., and Holmes, P. J. 1983a. A periodically forced piecewise linear oscillator. *J. Sound Vib.* **90**:129–155.

Shaw, S. W., and Holmes, P. J. 1983b. Periodically forced impact oscillator with large dissipation. *Phys. Rev. Lett.* **51**:623–626.

Singh, A., and Joseph, D. D. 1989. Autoregressive methods for chaos on binary sequences for the Lorenz attract. *Phys. Lett. A* **135**:247–251.

Stronge, W. J. 1990. Rigid body collisions with friction. *Proc. R. Soc. London A* **431**:169–181.

Takens, F. 1981. Detecting strange attractors in turbulence. In Rand, D. A., and Young, L. S., eds., *Dynamical Systems and Turbulence*. Lecture Notes in Mathematics, No. 898. Springer-Verlag, New York, pp. 266–281.

Thompson, J. M. T., and Ghaffari, R. 1982. Chaos after period-doubling in the resonance of an impact oscillator. *Phys. Lett. A* **91**:5–8.

Thompson, M. G., Bishop, S. R., and Foale, S. 1994. An experimental study of low velocity impacts. *Mach. Vib.* **3**:10–17.

Tworzydlo, W., Becker, E., and Oden, J. 1992. Numerical modeling of friction-induced vibrations and dynamic instabilities. In Ibrahim, R. A., and Soom, A., eds., *Friction-Induced Vibration, Chatter, Squeal, and Chaos*, ASME DE-Vol. 49, 13–32.

Whiston, G. S. 1992. Singularities in vibro-impact systems. *J. Sound Vib.* **152**:427–460.

Wiggins, S. 1990. *Introduction to Applied Nonlinear Dynamical Systems and Chaos*. Springer-Verlag, New York.

10

NONLINEAR DYNAMICS
AND FRACTURE

M. MARDER

10.1 INTRODUCTION

The study of nonlinear dynamics is frequently associated with simple flows and maps. Simple equations are capable of displaying extremely complex behavior, leading to the thought that whenever complex behavior is observed, it can be described by simple maps. Unfortunately, this last conclusion is incorrect. Most phenomena in nature are simply too complicated to be reduced usefully just to a few degrees of freedom.

Nonlinear dynamics does not completely lose its power when faced with systems that require more than a one-dimensional map. It suggests a systematic way of approaching complicated systems that is of great value both in theoretical and experimental investigations.

1. *Scaling arguments*, relying on conservation laws, can provide valuable information even in the most complicated systems.

2. A system sometimes exhibiting complicated behavior in space and time should be studied in a *systematic* way around the point where it makes the transition from simplicity to complexity. Once the complexity is fully developed, systematic study is much more difficult.

3. In the neighborhood of the transition region, one has a greatly increased possibility of studying the system through *simple models*.

4. Simple models allow one to gain qualitative understanding. In many cases the qualitative phenomena encountered in the simplest cases are

Dynamics and Chaos in Manufacturing Processes, Edited by Francis C. Moon.
ISBN 0-471-15293-5 © 1998 John Wiley & Sons, Inc.

preserved even in more realistic settings, in which case the phenomena are said to be *universal*.

5. With the aid of understanding gained from ideal models, one may be able to relate observations to *bifurcation sequences* or low-dimensional models.

The study of brittle fracture provides an example of all these ideas.

1. A scaling argument, due to Mott (1948) and Dulaney and Brace (1960), permits one to identify the basic physical processes operating in dynamic fracture.

2. Systematic experimental study of brittle crack motion, focusing upon detailed records of dynamics, has provided a picture of dynamic instabilities in the propagation of fast cracks.

3. Dynamics of brittle solids may fruitfully be investigated in the context of an ideal brittle atomic solid.

4. This simple model introduces two qualitative ideas: that crack motion is forbidden in a range beginning at zero and proceeding up to about 20–30% of the Rayleigh wave speed and that the crack tip becomes unstable to a microcracking instability at about 50% of the Rayleigh wave speed. Real brittle materials such as glass and Plexiglas may well be in this universality class.

5. The onset of fracture turns out to be a somewhat unconventional subcritical bifurcation, the onset of microcracking resembles the scenario known as intermittency, and the motion of cracks in certain stable loading configurations can be reduced to a simple one-dimensional problem.

10.2 SCALING ARGUMENT

The subject of dynamic fracture initiated with a scaling argument due to Mott (1948), and then improved slightly by Dulaney and Brace (1960). Although technically superseded by careful mathematical studies of dynamic fracture, these simple arguments are very effective in explaining the basic forces at work during the fracture process, and they provide a good introduction to the major outstanding problems.

Consider a crack of length $l(t)$ growing at rate $v(t)$ in a plate under stress σ_∞ far from the crack, as shown in Figure 10.1. When the crack extends, its faces separate, causing the plate to relax within a circular region centered on the middle of the crack and with diameter of order l. The kinetic energy involved in moving a region of this size is guessed to be of the form

$$K = c_K l^2 v^2, \tag{1}$$

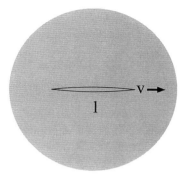

FIGURE 10.1. As a crack of length l expands at velocity v in an infinite plate, it disturbs the surrounding medium up to a distance on the order of l.

and the potential energy gained in releasing stress from the region is guessed to be of the form

$$P = -c_P l^2. \tag{2}$$

These guesses are correct for slowly moving cracks, but they fail qualitatively as the crack velocity approaches the speed of sound, in which case both kinetic and potential energies diverge. Fortunately, in this scaling argument, it is only the ratio of kinetic and potential energies that appears, and since both diverge in the same way, the error cancels out. The final process contributing to the energy balance equation is the creation of new crack surfaces, which takes energy $2\gamma l$, where γ is the fracture energy. So the total energy of the system containing a crack is given by

$$E = c_K l^2 v^2 + E_{qs}(l), \tag{3}$$

with

$$E_{qs}(l) = -c_P l^2 + 2\gamma l. \tag{4}$$

Consider first the problem of quasi-static crack propagation. If a crack moves forward only slowly, its kinetic energy will be negligible, and only the quasi-static part of the energy, E_{qs}, will be important. This function is graphed in Figure 10.2. It costs energy for very short cracks to elongate, and in fact such cracks would heal and travel backward if it were not for irreversible processes, such as oxidation or reconstruction of the crack surface, that typically prevent this from happening. That the crack grows at all is due to additional irreversible processes, sometimes chemical attack on the crack tip, sometimes vibration or other irregular mechanical stress. It should be emphasized that the system energy E increases as a result of these processes.

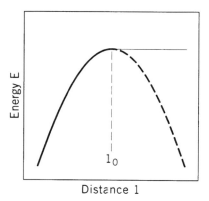

FIGURE 10.2. Energy of a plate with a crack as a function of length. In the first part of its history, the crack grows quasi-statically, and its energy increases. At l_0 the crack begins to move rapidly, and energy is conserved.

Eventually, at length l_0, the energy gained by relieving elastic stresses in the body exceeds the cost of creating new surface, and the crack becomes able to extend spontaneously. One sees that at l_0 the energy functional $E_{qs}(l)$ has a quadratic maximum, so Eq. (4) can be rewritten

$$E_{qs}(l) = E_{qs}(l_0) - c_P(i - l_0)^2, \qquad l_0 = \frac{\gamma}{c_P}. \tag{5}$$

The problem of fracture initiation boils down to calculating l_0, given things such as external stresses which in the present case have all been condensed into the constant c_P. Dynamic fracture starts in the next instant, and because it is so rapid, the energy of the system is conserved, remaining at $E_{qs}(l_0)$. Using Equations (3) and (5), with $E = E_{qs}(l_0)$, gives

$$v(t) = \sqrt{\frac{c_P}{c_K}} \left(1 - \frac{l_0}{l}\right) = v_{max}\left(1 - \frac{l_0}{l}\right). \tag{6}$$

This equation predicts that the crack will accelerate until it approaches the speed v_{max}. The maximum speed cannot be deduced from these arguments, but Stroh (1957) correctly argued that v_{max} should be the Rayleigh wave speed, the speed at which sound travels over a free surface. One needs only to know the length at which a crack begins to propagate in order to predict all the following dynamics.

Although the forms of kinetic and potential energy, Equations (1) and (2), are wrong as a crack approaches the Rayleigh wave speed, there is the remarkable result, due to Kostrov (1974), Eshelby (1969), Freund 1972a, 1972b, 1972c, 1973, 1974, 1990), and Willis (1990), that for a semi-infinite crack in an infinite two-dimensional plate, with a constant force pushing the faces of the crack apart, an exact solution of the partial differential equations for elasticity gives a complicated analytical expression, which differs from Equation (6) by at most a few percent. To be more precise, these calculations show that if one surrounds the crack tip with a small loop that moves with it, then the energy per unit crack extension flowing through the loop is

$$2\gamma = \Gamma \frac{l}{l_0}\left(1 - \frac{v}{v_R}\right), \tag{7}$$

where Γ is a constant with dimensions of energy per area, l_0 is the length of the crack at fracture initiation, and v_R is the Rayleigh wave speed. Solving Equation (7) for v, one immediately reproduces Equation (6).

10.3 SYSTEMATIC EXPERIMENTS

Experiments have never seemed to agree with Equation (6). One of the first experiments to test it quantitatively was by Kobayashi, Ohtani, and Sano (1974), whose data are reproduced in Figure 10.3. Similar data have now been

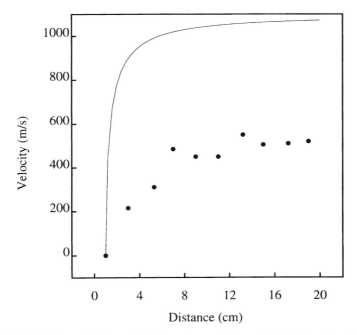

FIGURE 10.3. Data of Kobayashi, Ohtani, and Sano (1974) showing that cracks accelerate much more slowly than the scaling theory predicts.

duplicated many times in a variety of brittle amorphous materials, with similar results (Ravi-Chandar and Knauss, 1984).

However, brief observation of Equation (7) shows that there is a way out of the difficulty. One has only to suppose that the fracture energy γ is a function of velocity, and one obtains instead of Equation (6),

$$v = v_R \left(1 - \frac{2\gamma(v)l_0}{\Gamma l} \right). \tag{8}$$

By allowing the possibility that the fracture energy $\gamma(v)$ increases rapidly as velocity increases, one can obtain almost any profile of velocity versus crack length desired. Taking this idea further, one can view Equation (7) as a way to extract the velocity dependence of fracture energy from experimental velocity measurements. For this reason, study of the nonlinear dynamics of fracture at the University of Texas began with the aim of recording crack velocities with great precision, at high speed, and with great detail (Fineberg et al., 1991, 1992). We developed a technique that permits measurement of crack velocities to within ± 10 m/s at 20 MHz. Samples of crack speeds measured in this way appear in Figure 10.4.

Two main conclusions were drawn from these experiments. First, at a velocity around one-third of the Rayleigh wave speed, the motion of a crack in a brittle amorphous solid becomes unstable to an instability that causes the velocity to oscillate violently at a frequency of around 600 kHz. Second, steady crack propagation seems to be impossible between speeds of around 1 m/s and 180 m/s. We devoted some effort to discovering whether the onset of the velocity oscillations can be described by a standard type of bifurcation, for example, as a Hopf bifurcation (Fineberg et al., 1991). This approach was not successful and indicated that rapid fracture cannot be reduced to simple low-dimensional models. Thus we were required to develop simple models of a type that was more appropriate. They displayed a surprising resemblance to the experiment.

10.4   SIMPLE MODELS

One description of a macroscopic brittle solid is that it displays no nonlinearity in the stress-strain curve up to the instant of failure. The ideal brittle solid takes this idea down to the microscopic level. I define the ideal brittle solid to be a collection of mass points connected to nearest neighbors by perfectly linear forces. However, when the extension between each mass and its neighbor exceeds some critical extension, the force drops immediately to zero.

In the context of fracture, static models of this type were first examined by Thomson, Hsieh, and Rana (1971), and their dynamic behavior was first examined by Slepyan (1981). Since a lengthy technical examination of these

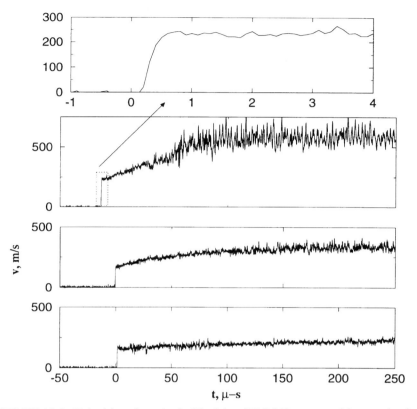

FIGURE 10.4. Velocities of cracks in Plexiglas (PMMA) measured by monitoring the resistance of a very thin aluminum coating at 20 MHz. By preparing plates with initial notches of different sizes, different final speeds can be obtained. The uppermost figure shows a detail of crack initiation, while second from the top is a graph of the crack accelerating to the point at which it exhibits a dynamic instability, appearing as large oscillations in the velocity.

models has recently been published (Marder and Gross, 1995), I would like just to focus on the physical reasons that these models behave differently than one might expect. For detailed mathematical justification for all the claims that follow, see Marder and Gross (1995).

The simplest possible model for an ideal brittle solid is one-dimensional. It is depicted in Figure 10.5 and described mathematically by

$$\ddot{x}_i = \begin{cases} x_{i+1} - 2x_i + x_{i-1} & \text{coupled to neighbors,} \\ +(1/N)(\Delta\sqrt{2N+1} - x_i) & \text{driving term,} \\ -2x_i\theta(1 - x_i) & \text{bonds that snap,} \\ -b\dot{x}_i & \text{dissipation.} \end{cases} \tag{9}$$

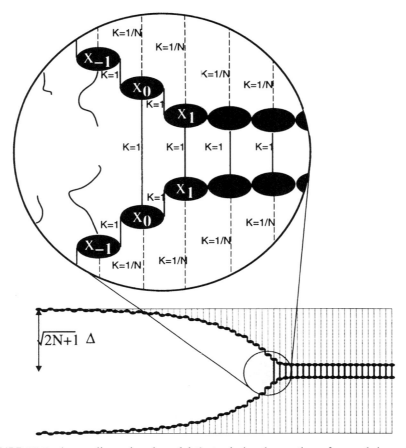

FIGURE 10.5. A one-dimensional model that mimics the motion of a crack in a strip, incorporating effects of discreteness. One can view it as a model for the atoms lying just along the surface of a crack. The mass points are only allowed to move vertically and are tied to their vertical neighbors with springs that break when they exceed unit extension. The lower portion of the figure shows an actual steadily moving solution of the model with velocity $v = 0.5$. Only cases where the mass points move perfectly symmetrically about the crack line will be considered; Equation (9) does not treat the upper and lower masses as separate degrees of freedom.

A number of the symbols above require discussion.

N

One can think of Equation (9) as describing a collection of atoms that lies just along a crack line, within a macroscopic strip of brittle material. Above and below are N additional rows of atoms, where N might be large, on the order of 10^9. However, in the interest of simplicity, in this one-dimensional model one chooses not to model these additional rows of atoms as explicit degrees of freedom and makes all their atoms massless. When N rows of atoms pull on the atoms on the crack line, their effective force constant is $1/N$ times the force constant between neighboring atoms; hence the factor of $1/N$ on the right-hand side of Equation (9).

Δ

The displacement Δ is a dimensionless measure of the external strain placed upon the strip. Far to the left of the crack tip, atoms rise to the equilibrium height $\sqrt{2N+1}\Delta$, and far to the right of the tip, atoms begin at the equilibrium height $\Delta/\sqrt{2N+1}$. One can check that the energy stored in the strip per bond far to the right is $2\Delta^2$ and that the energy needed to bring each bond to the snapping point is precisely 2. Therefore no matter how large N may be, $\Delta = 1$ defines the critical strain at which, in principle, one has enough energy stored to the right of the crack to allow it to propagate; $\Delta = 1$ should correspond to the Griffith point or to the critical stress intensity factor.

b

One must include a small amount of dissipation through the coefficient b. The need for this term is largely formal; without it the mathematics do not know which direction time is supposed to run. It is perfectly sensible to work in the limit $b \rightarrow 0$ but impossible to set $b = 0$ at the outset.

The most compact summary of analytical results for the model defined by Equation (9) is contained in Figure 10.6. This figure displays the velocity of steady state crack motion as a function of external strain Δ for $N = 100$ and

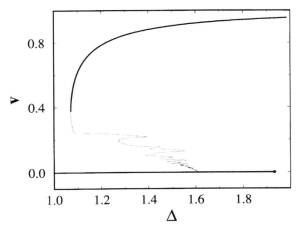

FIGURE 10.6. Velocity of a crack v plotted as a function of the driving force Δ for the model defined by Equation (9). The calculation is carried out for $N = 100$ and $b \rightarrow 0$. The thick upper line indicates physically realizable solutions, and the line along $v = 0$ indicates a range of lattice-trapped solutions.

in the limit $b \rightarrow 0$. There are four remarkable things about the figure:

1. Stationary cracks are stable over a wide range of Δ. This phenomenon is known as lattice trapping and is discussed extensively by Thomson (1986).

2. A branch of rapidly moving cracks is also stable and overlaps much of the lattice-trapping regime. For $N = 100$ and $b \rightarrow 0$, stable moving cracks first appear for $\Delta = 1.05$ and are only 5% greater than the Griffith value $\Delta = 1$.

3. Even in the limit $b \rightarrow 0$, moving cracks require more energy than is available at $\Delta = 1$ in order to propagate, and this energy increases as velocity v increases. The moving crack spontaneously puts this energy into high-frequency waves, evident in Figure 10.5, as emphasized by Slepyan (1981). In the limit $b \rightarrow 0$, the crack tip becomes a source of high-frequency radiation, so the assumption in fracture mechanics that all energy moves in to the crack tip fails. However, in real physical systems, these waves thermalize within the process zone around the crack tip, and manifest themselves as heat. In this sense fracture of an ideal brittle solid necessarily involves a large amount of dissipation.

4. There is a range of velocities, from around $v = 0$ to $v = 0.3$, in which steady state crack motion is impossible. Cracks can achieve these speeds temporarily while accelerating up to higher levels but cannot propagate stably in this forbidden band.

The last two points both follow from general features of the dispersion relation for moving objects in lattices. The way that the crack excites moving waves is described by the following general argument:

Consider the motion of a particle through a lattice whose particles are loc − ated at $\mathbf{R}_i$, and in which phonons are described by the dispersion relation

$$\omega_\alpha(\mathbf{k}). \tag{10}$$

If the particle moves with constant velocity $\mathbf{v}$, and interacts with the various ions according to some function $\mathscr{I}$, then the system can be described by

$$m\ddot{u}_\mu^l = -\sum_{vl'} D_{\mu v}(\mathbf{R}_l - \mathbf{R}_{l'})u_v^{l'} + \mathscr{I}_\mu(\mathbf{R}_l - \mathbf{v}t). \tag{11}$$

Multiplying everywhere by $e^{i\mathbf{k}\cdot\mathbf{R}_l}$ and summing over l gives

$$m\ddot{u}_\mu(\mathbf{k}) = \sum_v D_{\mu v}u_v(\mathbf{k}) + e^{i\mathbf{k}\cdot\mathbf{v}t}\mathscr{I}_\mu(\mathbf{k}). \tag{12}$$

Inspection shows that the lattice frequencies ω which are excited in this way are those for which

$$\omega(\mathbf{k}) = \mathbf{v} \cdot \mathbf{k}. \tag{13}$$

In a continuum, Equation (13) can only obtain when $\mathbf{k} = 0$, or when v equals the sound speed c. However, as shown in Figure 10.7, in a lattice, there is always at least one solution of Equation (13) for any $v < c$, and as v decreases, the number of solutions of Equation (13) increases. By analyzing the solutions of Equation (13), one can predict the frequencies of the waves left behind by the crack. However, more thorough analysis is needed in order to determine the amplitudes of the waves.

As discussed in the caption to Figure 10.7, cracks traveling faster than $v \sim 0.4$ do not send any traveling waves out ahead of the tip. At velocities lower than this, they begin to do so. Rapidly the amplitude of high-frequency waves ahead of the tip grows, until the lattice is unable to sustain them; in a sense the emitted radiation melts the lattice and destroys the solutions. If one supposes the crack tip to be at any particular location, and then investigates the vibrations ahead of the tip, one finds that these vibrations are of such an amplitude that the bonds ahead of the tip must already have broken. The only way to resolve this inconsistency is to realize that crack motion at such velocities is impossible, and numerical simulation verifies that in fact no such solutions exist.

The study of two-dimensional ideal brittle solids is formally nearly identical to the study of the one-dimensional model, and only one qualitatively new phenomenon emerges. This phenomenon is important, however, because it

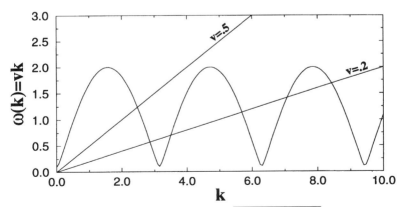

FIGURE 10.7. The dispersion relation $\omega(k) = \sqrt{1/N + 4\sin^2(k/2)}$ for the left-hand side of the model described by Equation (9), plotted versus k, and compared with vk. There is always at least one crossing corresponding to a wave traveling to the left, and at all crack velocities v, the crack must leave vibrations at this frequency in its wake. To the right of the crack, the dispersion relation is $\omega(k) = \sqrt{1/N + 4 + 4\sin^2(k/2)}$. For $v > 0.4$, the solutions of $\omega = vk$ for this dispersion relation also correspond to waves moving toward the left, and above this crack velocity, the crack sends no waves ahead of itself.

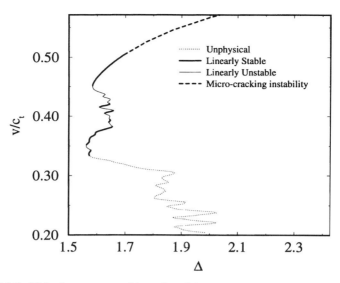

FIGURE 10.8. Velocity v measured in units of the transverse wave speed c_t as a function of Δ, for a two-dimensional mode I ideal brittle solid. The thick lines are physically realizable states, the thin solid lines are linearly unstable, and the dotted lines are unphysical states.

provides a possible resolution for the terminal speed problem. One can see it displayed in Figure 10.9. As the crack exceeds a critical velocity, which for a perfect triangular lattice is around $v = 0.66$, instead of simply breaking the bonds ahead of the crack tip, the crack begins occasionally to snap bonds off to the side as well. As one increases Δ to drive the crack harder, the density of bonds broken to the side of the crack increases, and eventually complicated dendritic structures emerge.

10.5 UNIVERSALITY

With predictions in hand from an ideal model, one can now ask whether the basic qualitative phenomena uncovered there have any generality. There are two ways to pose this question. First, one can compare the predictions with experiment. Second, one can carry out numerical work on models that are no longer susceptible to analytical solution, comparing them to the exact results.

Comparison with experiment is so far quite promising, although not entirely conclusive. In Figure 10.4, showing results in Plexiglas, it is evident that steady crack motion does not occur in a range of velocities up to around 20% of the Rayleigh wave speed. We have obtained similar results in soda-lime glass, although the story there is complicated by the difficulties of obtaining a perfectly sharp initial crack. The instability resulting in velocity oscillations does in fact correspond to a micro-cracking instability, as shown in Figure

Δ=1.147
v/c=0.645

Δ=1.165
v/c=0.630

Δ=1.376
v/c=0.624

Δ=1.835
v/c=0.775

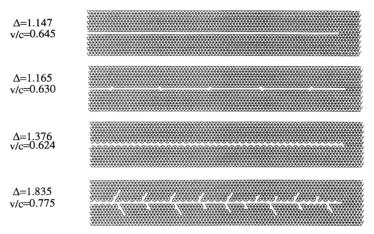

FIGURE 10.9. Pictures of broken bonds left behind the crack tip at four different values of Δ, from simulations of Liu (1993). The top pattern shows simple bonds broken by a steady state crack. At a value of Δ slightly above the critical one where horizontal bonds occasionally snap, the pattern is periodic. All velocities are measured relative to the sound speed $c = \sqrt{3}/2$. Notice that the average velocity can decrease relative to the steady state, although the external strain has increased. As the strain Δ increases further, other periodic states can be found, and finally states with complicated spatial structure. The simulations are carried out in a strip with half-width $N = 9$, length 200, and $b = 0.01$. The front and back ends of the strip have short energy-absorbing regions to damp traveling waves.

10.10. Thus, from the experimental side, there is reason to hope that the ideal brittle solid exhibits a type of fracture with a rather broad universality class.

Numerical simulations now in progress support this same view. The simulations change the ideal brittle solid by changing the form of the interatomic interaction, making the lattice more or less random, and by including effects of temperature. A mode III crack moving in a lattice where the bond strengths vary randomly by 20% from site to site appears in Figure 10.11, and velocity versus driving force in the presence of various degrees of randomness is shown in Figure 10.12. All the basic phenomena uncovered in the analytical work remain present in numerical studies of more complicated models, for mode III. However, there are some early indications that for mode I fracture, the situation will be more complicated and is not yet ready for discussion.

10.6 BIFURCATIONS AND LOW-DIMENSIONAL DYNAMICS

Some of the conclusions for ideal brittle solids are profitably discussed in terms of simple ideas from bifurcation theory. The way that crack initiation occurs overturns the conceptual view of fracture initiation that has held since Griffith (1921).

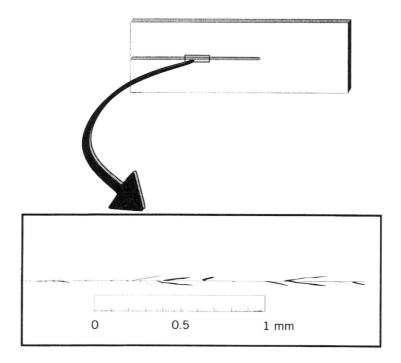

FIGURE 10.10. Drawing of small microcracks that begin to emerge from a crack after it passes a velocity of around 330 m/s in Plexiglas (PMMA). The basic geometry is the same as that shown in Figure 10.9, with a spacing on the order of 0.07 mm. This diagram shows the microcracks near the point where they first appear; later they are longer and more densely packed. The microcracks shown are those that appear in a microscope whose depth of field is about 100 μ, and the pattern of microcracks changes as one looks at different vertical planes by changing the focus of the microscope.

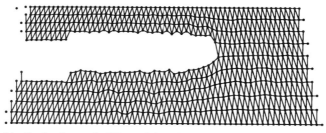

FIGURE 10.11. Path of a mode III crack in a lattice where bond strengths vary by 20% from site to site. In addition the interatomic interaction is no longer purely brittle but curves over and has a long flat region before dropping to zero.

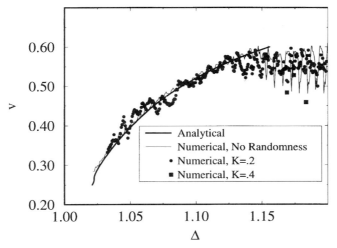

FIGURE 10.12. Velocity versus driving force for a mode III crack in a lattice where bond strengths vary randomly, compared with analytical results for the ideal brittle solid. When bond strengths vary at the 20% level, effects on the crack motion are insignificant, but at the 40% level the crack encounters a strong pinning site and stops at relatively high stresses. Such pinning is much easier in two dimensions than in three, and two-dimensional modeling is no longer adequate.

The traditional view, shown in Figure 10.13, is that fracture begins at a critical external stress, or a critical stress intensity factor, the Griffith point. Thus fracture is supposed to be a supercritical bifurcation. By contrast, for the ideal brittle solid, the Griffith point does not exist; fracture is subcritical. A stationary crack must somehow jump over an activation barrier in order to begin to run, and in a system with any level of noise, this jump will certainly occur before Δ rises to the end of the lattice-trapping regime. It must be emphasized that the place where the jump occurs in inherently unpredictable, although one can ask about such things as the mean transition time at a given temperature and load.

There is a way to drive crack motion so that a very simple one-dimensional model becomes applicable. Suppose that one loads a crack stably, and then insists on dragging the crack along at a v speed which lies in the range forbidden for steady state cracks. Such loading is indicated schematically in Figure 10.14, and can be achieved in the laboratory by a number of devices, for example, by lowering a heated sample into a cold bath at a fixed rate. Since the crack cannot move steadily at the driving velocity, it waits for the loading to reach a level at which it becomes unstable. It then races ahead at high speed until it exits the region, where the loading allows it to continue to propagate, and drops rapidly to zero velocity. This behavior is pictured in Figure 10.15. It has not yet been investigated experimentally.

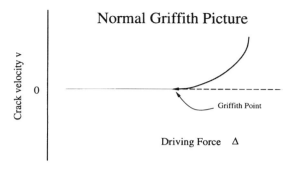

Normal Griffith Picture

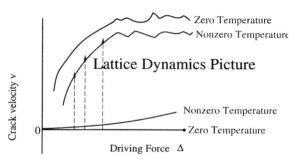

Lattice Dynamics Picture

FIGURE 10.13. The traditional view of fracture initiation is that crack velocity versus external loading resembles the top sketch. A crack is stationary until the Griffith point, when it begins to move. When effects of temperature or environmental degradation are taken into account, rather than remaining completely stationary below the Griffith point, the crack might move exponentially slowly. However, the ideal brittle solid behaves differently. Slow crack growth and rapid fracture constitute two completely separate types of motion and the two can coexist for certain ranges of external conditions.

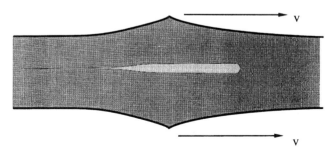

FIGURE 10.14. A sample is loaded in such a way as to force a crack to move at an externally imposed velocity. If the crack drops back, the loading grows so much that it is forced to go forward, and if the crack goes too far, the loading drops to zero and the crack must stop.

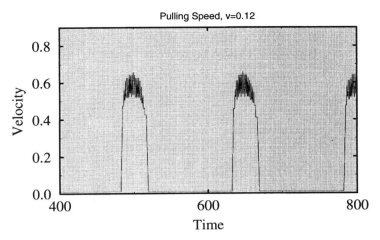

FIGURE 10.15. When forced to move at a velocity that lies in its forbidden band, a crack undergoes stick-slip motion. This figure was prepared by Eric Gerde.

10.7 CONCLUSIONS

This chapter has shown that the methodology of nonlinear dynamics can help develop a new view of dynamic fracture, demonstrating that careful attention to dynamical behavior combined with simple modeling can uncover new phenomena and help explain them.

The model systems discussed in this chapter offer a picture of what ideal brittleness means. Absence of dissipation is not central to brittleness, since a model with no apparent dissipation develops its own spontaneously. The most important feature of such a model is that it demands the transition between static and rapid crack motion be rapid and unpredictable. A brittle material is one with a band of forbidden velocities. It is this forbidden band that allows the models to mimic features of brittleness, so common in the everyday world but not before captured in mathematical calculations.

ACKNOWLEDGMENTS

I am grateful for financial support from the Alcoa Foundation, and for travel funds from the U.S., Israel Binational Science Foundation, Grant 920-00148/1.

REFERENCES

Bell, J. F. 1984. *The Experimental Foundations of Solid Mechanics.* Springer-Verlag, Berlin.

Dulaney, E. N., and Brace, W. F. 1960. Velocity behavior of a moving crack. *J. Appl. Phys.* **31**:2233–2236.

Eshelby, J. D. 1969. The elastic field of a crack extending nonuniformly under general anti-plane loading. *J. Mech. Phys. Sol.* **17**:1–99.

Fineberg, J., Gross, S., Marder, M., and Swinney, H. 1991. Instability in dynamic fracture. *Phys. Rev. Lett.* **67**:457–460.

Fineberg, J., Gross, S., Marder, M., and Swinney, H. 1992. Instability in the propagation of fast cracks. *Phys. Rev. B* **45**:5146–5154.

Freund, L. B. 1972a. Energy flux into the tip of an extending crack in an elastic solid. *J. Elast.* **2**:341–349.

Freund, L. B. 1972b. Crack propagation in an elastic solid subjected to general loading. I: Constant rate of extension. *J. Mech. Phys. Sol.* **20**:129–140.

Freund, L. B. 1972c. Crack propagation in an elastic solid subjected to general loading. II: Nonuniform rate of extension. *J. Mech. Phys. Sol.* **20**:141–152.

Freund, L. B. 1973. Crack propagation in an elastic solid subjected to general loading. III: Stress wave loading. *J. Mech. Phys. Sol.* **21**:47–61.

Freund, L.B. 1974. Crack propagation in an elastic solid subjected to general loading. IV: Obliquely incident stress pulse. *J. Mech. Phys. Sol.* **22**:137–146.

Freund, L. B. 1990. *Dynamic Fracture Mechanics.* Cambridge University Press, New York.

Griffith, A. A. 1920. The phenomenon of rupture and flow in solids. *Phil. Trans. R. Soc. London A* **221**:163–198.

Kanninen, M. F., and Popelar, C. 1985. *Advanced Fracture Mechanics.* Oxford University Press, New York.

Kobayashi, A., Ohtani, N., and Sato, T. 1974. Phenomenological aspects of viscoelastic crack propagation. *J. Appl. Polymer Sci.* **18**:1625–1638.

Kostrov, B. V. 1975. On the crack propagation with variable velocity. *Int. J. Fracture* **11**:4–56.

Liu, X. 1993. Dynamics of fracture propagation. Ph.D. Dissertation. University of Texas at Austin.

Marder, M., and Gross, S. 1995. Origin of crack tip instabilities. *J. Mech. Phys. Sol.* **43**:1–48.

Mott, N. F. 1948. Brittle fracture in mild steel plates. *Eng.* **165**:16–18.

Ravi-Chandar, K., and Knauss, W. G. 1984a. An experimental investigation into dynamic fracture. I: Crack initiation and arrest. *Int. J. Fracture* **20**:209–222.

Ravi-Chandar, K., and Knauss, W. G. 1984b. An experimental investigation into dynamic fracture. II: Microstructural aspects. *Int. J. Fracture* **25**:247–262.

Ravi-Chandar, K., and Knauss, W. G. 1984c. An experimental investigation into dynamic fracture. III: On steady state crack propagation and branching. *Int. J. Fracture* **26**:141–154.

Ravi-Chandar, K., and Knauss, W. G. 1984d. An experimental investigation into dynamic fracture. IV: On the interaction of stress waves with propagating cracks. *Int. J. Fracture* **266**:189–200.

Slepyan, L. I. 1981. Dynamic of brittle fracture in lattice. *Sov. Phys. Dokl.* **26**:538–540.

Stroh, A. N. 1957. A theory of the fracture of metals. *Adv. Phys.* **6**:418–465.

Thomson, R. 1986. Physics of fracture. *Sol. State Phys.* **39**:1–129.

Thomson, R., Hsieh, C., and Rana, V. 1971. Lattice trapping of fracture cracks. *J. Appl. Phys.* **42**:3154.

Willis, J. R. 1990. Accelerating cracks and related problems. In Eason, G, ed. *Elasticity: Mathematical Methods and Applications.* Halston, New York, p. 397.

Yoffe, E. H. 1951. The moving Griffith crack. *Phil. Mag.* **42**:739–750.

11

CONTROLLING CHAOS
IN MECHANICAL
SYSTEMS

ERNEST BARRETO, YING-CHENG LAI,
and CELSO GREBOGI

11.1 INTRODUCTION

The control of chaos by unstable periodic orbits embedded in a chaotic attractor was first proosed in 1990 (Ott, Grebogi, and Yorke, 1990). Numerous experiments in many different fields have demonstrated the feasibility of this approach. This technique has been applied for example to mechanical systems (Ditto, Rauseo, and Spano, 1990; Hübinger et al., 1994; Starrett and Tagg, 1994; Moon, Johnson, and Holmes, 1996), lasers (Roy et al., 1992; Gills et al., 1992; Bielawski, et al., 1993, 1994; Reyl et al., 1993; Meucci et al., 1994), circuits (Hunt, 1991; Johnson and Hunt, 1993; Gauthier et al., 1994), chemical reactions (Petrov et al., 1993, 1994), biological systems (Schiff et al., 1994; Garfinkel et al., 1992), communication technology (Hayes, Grebogi, and Ott, 1993; Hayes et al., 1994), and energy production methods (Rhode et al., 1995; Daw et al., 1995). Furthermore it is possible to switch efficiently from one unstable periodic orbit to another at will (Ditto, Rauseo, and Spano, 1990; Romeiras et al., 1992; Kostelich et al., 1993; Barreto et al., 1995a). The basic idea is as follows: First, one chooses an unstable periodic orbit embedded in an attractor that yields the best system performance according to some criteria. Second one allows the trajectory to enter a small region around the desired

Dynamics and Chaos in Manufacturing Processes, Edited by Francis C. Moon.
ISBN 0-471-15293-5 © 1998 John Wiley & Sons, Inc.

orbit, which is bound to happen because of the ergodicity of the chaotic attractor. As this occurs, small, judiciously chosen parameter perturbations are applied to force the trajectory to approach the unstable periodic orbit. This method is extremely flexible because chaotic attractors typically have embedded within them an infinite number of unstable periodic orbits. Thus, by choosing a set of system parameters that yields chaos, one has access to a great many different system behaviors.

In this chapter, we illustrate the idea of controlling chaos using both low- and high-dimensional dynamical systems. As a pedagogic example, in Section 11.2 we use the logistic map to elucidate the fundamental concepts and methodology. In Section 11.3 we emphasize the geometrically intuitive two-dimensional case and present a general algorithm applicable to systems of any dimension. In Section 11.4 we present an application to a mechanical system, the kicked double rotor. We also discuss a method for rapidly steering chaotic trajectories to a desirable periodic orbit on the attractor. We draw our conclusions in Section 11.5.

11.2   A ONE-DIMENSIONAL EXAMPLE

The basic idea behind controlling chaos can be understood by considering a simple model system. We consider one of the best understood chaotic systems, the simple one-dimensional logistic map:

$$x_{n+1} = f(x_n, \lambda) = \lambda x_n (1 - x_n), \tag{1}$$

where x is restricted to the unit interval $[0, 1]$ and λ is a control parameter. It is known that this map develops chaos via the period-doubling bifurcation route (Feigenbaum, 1978). For $0 < \lambda < 1$, the asymptotic state of the map (or the attractor of the map) is $x = 0$; for $1 < \lambda < 3$, the attractor is a nonzero fixed point $x_F = 1 - 1/\lambda$; for $3 < \lambda < 1 + \sqrt{6}$, this fixed point is unstable and the attractor is a stable period-2 orbit. As λ is increased further, a sequence of period-doubling bifurcations occurs in which successive period-doubled orbits become stable. The period-doubling cascade accumulates at $\lambda = \lambda_\infty \approx 3.57$, after which chaos arises.

Consider the case $\lambda = 3.8$ shown in Figure 11.1a where the system is apparently chaotic. An important characteristic of a chaotic attractor is that there exists *an infinite number of unstable periodic orbits embedded within it*. Shown in the figure are, for example, a fixed point $x_F \approx 0.7368$ and a period-2 orbit with components $x(1) \approx 0.3737$ and $x(2) \approx 0.8894$, where $x(1) = f(x(2))$ and $x(2) = f(x(1))$.

Suppose that we want to avoid chaos at $\lambda = 3.8$. In particular, we want trajectories resulting from a randomly chosen initial condition x_0 to be as close as possible to the period-2 orbit shown in Figure 11.1a, assuming that this

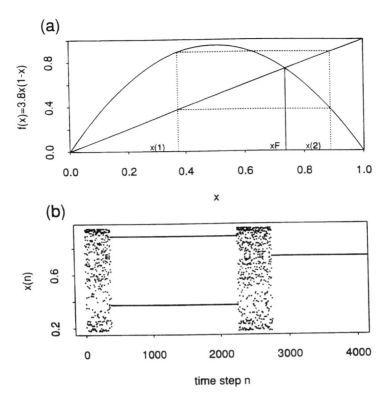

FIGURE 11.1. (*a*) The logistic map $x_{n+1} = f(x_n) = 3.8x_n(1 - x_n)$. An unstable fixed point and an unstable period-2 orbit are also shown. (*b*) Time series illustrating the control of the period-2 orbit and the fixed point. The chaotic trajectory begins from $x_0 = 0.28$. At $n = 381$, the trajectory falls in an ε-neighborhood of the period-2 orbit, after which the parameter control is turned on to stabilize the trajectory around the period-2 orbit. At $n = 2200$, the control is turned off. At $n = 2757$, the chaotic trajectory comes close to the fixed point and is controlled in subsequent iterations. We choose $\varepsilon = 10^{-3}$. The maximum allowed parameter perturbation is $\delta = 5 \times 10^{-3}$.

period-2 orbit gives the best system performance. Of course we can choose the desired asymptotic state of the map to be any of the infinite number of unstable periodic orbits. Suppose further that the parameter λ can be fine-tuned in a small range around the value $\lambda_0 = 3.8$, namely we allow λ to vary in the range $[\lambda_0 - \delta, \lambda_0 + \delta]$, where $\delta \ll 1$. Due to the nature of the chaotic attractor, a trajectory that begins from an arbitrary value of x_0 will fall, with probability one, into the neighborhood of the desired period-2 orbit at some later time. Because of the nature of chaos, the trajectory would diverge quickly from the period-2 orbit if we do not intervene. Our task is to program the variation of the control parameter so that the trajectory stays in the neighborhood of the period-2 orbit as long as the control is present. In general, the small parameter perturbations will be time dependent. We emphasize that it is important to apply only small parameter perturbations. If large parameter perturbations are allowed, then obviously we can eliminate chaos by varying λ from 3.8 to 2.0, for example. Such a large change is not interesting.

The logistic map in the neighborhood of a periodic orbit can be approximated by a linear equation expanded around the periodic orbit. Denote the target period-m orbit to be controlled as $x(i)$, $i = 1,\ldots, m$, where $x(i + 1) = f(x(i))$ and $x(m + 1) = x(1)$. Assume that at time n the trajectory falls into the neighborhood of component i of the period-m orbit. The linearized dynamics in the neighborhood of component $i + 1$ is then

$$x_{n+1} - x(i + 1) = \frac{\partial f}{\partial x}[x_n - x(i)] + \frac{\partial f}{\partial \lambda}\Delta\lambda_n$$

$$= \lambda_0[1 - 2x(i)][x_n - x(i)] + x(i)[1 - x(i)]\Delta\lambda_n, \qquad (2)$$

where the partial derivatives in (2) are evaluated at $x = x(i)$ and $\lambda = \lambda_0$. We require x_{n+1} to stay in the neighborhood of $x(i + 1)$. Hence we set $x_{n+1} - x(i + 1) = 0$, which gives

$$\Delta\lambda_n = \lambda_0 \frac{[2x(i) - 1][x_n - x(i)]}{x(i)[1 - x(i)]}. \qquad (3)$$

Equation (3) holds only when the trajectory x_n enters a small neighborhood of the period-m orbit, namely when $|x_n - x(i)| \ll 1$; hence the required parameter perturbation $\Delta\lambda_n$ is small. Let the length of a small interval defining the neighborhood around each component of the period-m orbit be 2ε. In general, the required maximum parameter perturbation δ is proportional to ε. Since ε can be chosen to be arbitrarily small, δ also can be made arbitrarily small. As we will see, the average transient time before a trajectory enters the neighborhood of the target periodic orbit depends on ε (or δ). When the trajectory is outside the neighborhood of the target periodic orbit, we do not apply any parameter perturbation, and the system evolves at its nominal parameter value λ_0. Hence we usually set $\Delta\lambda_n = 0$ when $\Delta\lambda_n > \delta$. Note that the parameter perturbation $\Delta\lambda_n$ depends on x_n and is time dependent.

The above strategy for controlling the orbit is very flexible for stabilizing different periodic orbits at different times. Suppose that we first stabilize a chaotic trajectory around the period-2 orbit shown in Figure 11.1a. Then we might wish to stabilize the fixed point in Figure 11.1a, assuming that the fixed point would correspond to a better system performance at a later time. To achieve this change of control, we simply turn off the parameter control with respect to the period-2 orbit. Without control, the trajectory will diverge from the period-2 orbit exponentially. We let the system evolve at the parameter value λ_0. Due to the nature of chaos, there comes a time when the chaotic trajectory enters a small neighborhood of the fixed point. At this time we turn on a new set of parameter perturbations calculated with respect to the fixed point. The trajectory can then be stabilized around the fixed point.

Figure 11.1b shows an example where we first control the period-2 orbit and then the fixed point shown in Figure 11.1a. The initial condition is $x_0 = 0.28$. At time $n = 381$, the trajectory enters the neighborhood of component $x(1)$ of the period-2 orbit. For subsequent iterations the parameter control calculated from (3) is used to stabilize the trajectory around the period-2 orbit. At time $n = 2200$, we chose to stabilize the trajectory around the fixed point, and hence we turn off the parameter perturbation. The trajectory quickly leaves the period-2 orbit and becomes chaotic. At time $n = 2757$, the trajectory falls into the neighborhood of the fixed point. Parameter perturbations calculated with respect to the fixed point are then turned on to stabilize the trajectory around the fixed point.

In the presence of external noise, a controlled trajectory will occasionally be "kicked" out of the neighborhood of the periodic orbit. If this behavior occurs, we turn off the parameter perturbation and let the system evolve by itself. With probability one the chaotic trajectory will enter the neighborhood of the target periodic orbit and be controlled again. This situation is illustrated in Figure 11.2a where we control the period-2 orbit. The noise is modeled by an additive term in the logistic map of the form $\eta\sigma_n$, where σ_n is a Gaussian distributed random variable with zero mean and unit standard deviation, and η is the noise amplitude. The effect of the noise is to turn a controlled periodic trajectory into an intermittent one in which chaotic phases (uncontrolled trajectories) are interspersed with laminar phases (controlled periodic trajectories). It is easy to verify that the averaged length of the laminar phase increases as the noise amplitude decreases, and the length tends to infinity as $\eta \to 0$.

Let us consider how many iterations are required on average for a chaotic trajectory originating from an arbitrarily chosen initial condition to enter the neighborhood ε of the target periodic orbit. Clearly the smaller the value of ε, the more iterations are required. In general, the average transient time $\tau(\varepsilon)$ before turning on control scales with ε as

$$\tau(\varepsilon) \sim \varepsilon^{-\gamma}, \tag{4}$$

where $\gamma > 0$ is a scaling exponent. For one-dimensional maps such as the logistic map, there usually exists a smooth probability density $\rho(x)$ for trajectory points on the attractor. The probability density ρ can be defined as the frequency that a chaotic trajectory visits a small neighborhood of the point x on the attractor. In such a case we have $\gamma = 1$, as can be seen by the following argument: The probability that a trajectory enters the neighborhood of a particular component (component i) of the periodic orbit is given by

$$P(\varepsilon) = \int_{x(i)-\varepsilon}^{x(i)+\varepsilon} \rho(x(i))dx \approx 2\varepsilon\rho(x(i)). \tag{5}$$

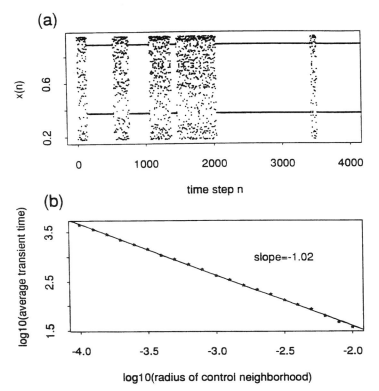

FIGURE 11.2. (*a*) The effect of additive noise modeled by $2.6 \times 10^{-4}\sigma_n$, where σ_n is a Gaussian random variable with zero mean and unit standard deviation. The noise can occasionally kick the controlled trajectory out of the neighborhood of the periodic orbit. (*b*) Log-log plot of the average time to achieve control $\tau(\varepsilon)$ versus ε, the size of the controlling neighborhood. Twenty values of ε are chosen on a logarithmic scale. For each ε, 2000 random initial conditions uniformly distributed in $[0, 1]$ are chosen to compute $\tau(\varepsilon)$. The scaling relation between $\tau(\varepsilon)$ and ε is well fitted by $\tau(\varepsilon) \sim \varepsilon^{-1}$.

Hence $\tau(\varepsilon) = 1/P(\varepsilon) \sim \varepsilon^{-1}$, and therefore $\gamma = 1$. This behavior is illustrated in Figure 11.2*b*, where $\tau(\varepsilon)$ is plotted on a logarithmic scale for the case of stabilizing the period-2 orbit in Figure 11.1*a*. Twenty values of ε were chosen in the range $[10^{-4}, 10^{-2}]$. For each ε we randomly choose 2000 initial conditions (with a uniform probability distribution) and calculate an average transient time. The slope of the straight line is approximately -1.02, indicating good agreement with the theoretical prediction of $\gamma = 1$. For higher-dimensional chaotic systems, the exponent γ can be related to the eigenvalues of the target periodic orbit (Ott, Grebogi, and Yorke, 1990).

A major advantage of the controlling chaos idea is that it can be applied to experimental systems in which a priori knowledge of the system is usually not known. A time series found by measuring one of the system's dynamical variables in conjunction with the time delay embedding method (Sauer, Casdagli, and Yorke, 1991; Ott, 1993), which transforms a scalar time series into a trajectory in phase space, is sufficient to determine the desired unstable periodic orbits to be controlled and the relevant quantities required to compute parameter perturbations (Ditto, Rauseo, and Spano, 1990; Garfinkel et al., 1992; Dressler and Nitsche, 1992; So and Ott, 1995). Another advantage of the method is its flexibility in choosing the desired periodic orbit to be controlled. The method has attracted growing interest in controlling dynamical systems and has been extended to higher-dimensional dynamical systems (Romeiras et al., 1992; Auerbach et al., 1992), Hamiltonian systems (Lai, Ding, and Grebogi, 1993a), the control of transient chaos (Tél, 1991) and chaotic scattering (Lai, Tél, and Grebogi, 1993c) and the synchronization of chaotic systems (Lai and Grebogi, 1993, 1994b). It also has been successfully implemented in various physical experiments (Ditto, Rauseo, and Spano, 1990; Garfinkel et al., 1992). In Section 11.3 we will describe the method formulated for two-dimensional maps.

11.3 CONTROLLING HIGHER-DIMENSIONAL SYSTEMS

The general algorithm for controlling chaos in higher-dimensional maps (or autonomous flows that can be reduced to maps on a Poincaré surface of section) can be formulated as follows. (By autonomous flow we mean that the vector field does not contain an explicit time dependence.) Consider the d-dimensional map,

$$\mathbf{X}_{n+1} = \mathbf{F}(\mathbf{X}_n, \mathbf{p}) \tag{6}$$

where $\mathbf{X}_n \in \mathbf{R}^d$, $\mathbf{F}$ is a smooth function of its variables, and $\mathbf{p} \in \mathbf{R}^r$ is a vector of r externally accessible control parameters. We restrict the parameter perturbations to be small, namely

$$|p^i - p_0^i| < \delta_i, \qquad i = 1, \ldots, r, \tag{7}$$

where $\mathbf{p}_0$ is some nominal parameter value and $\delta_i \ll 1$ defines the range of parameter variation. We wish to program the parameters $\mathbf{p}$ so that a chaotic trajectory is stabilized when it enters an ε neighborhood of the target periodic orbit. Let the desired period-m orbit be $\mathbf{X}(1, \mathbf{p}_0) \to \mathbf{X}(2, \mathbf{p}_0) \to \cdots \to \mathbf{X}(m, \mathbf{p}_0) \to \mathbf{X}(m + 1, \mathbf{p}_0) = \mathbf{X}(1, \mathbf{p}_0)$. The linearized dynamics in the neighborhood of component $i + 1$ of the period-m orbit is

$$\mathbf{X}_{n+1} - \mathbf{X}(i + 1, \mathbf{p}_0) = \mathbf{A} \cdot [\mathbf{X}_n - \mathbf{X}(i, \mathbf{p}_0)] + \mathbf{B}\Delta\mathbf{p}_n, \tag{8}$$

where $\Delta\mathbf{p}_n = \mathbf{p}_n - \mathbf{p}_0$, $\mathbf{A}$ is a $d \times d$ Jacobian matrix, and $\mathbf{B}$ is a $d \times r$ matrix:

$$\mathbf{A} = \mathbf{D}_x \mathbf{F}(\mathbf{X}, \mathbf{p})|_{\mathbf{X}(i), \mathbf{p}_0}, \tag{9}$$

$$\mathbf{B} = \mathbf{D}_p \mathbf{F}(\mathbf{X}, \mathbf{p})|_{\mathbf{X}(i), \mathbf{p}_0}.$$

Now writing $\Delta\mathbf{p}_n = -\mathbf{K} \cdot [\mathbf{X}_n - \mathbf{X}(i, \mathbf{p}_0)]$, we have

$$\mathbf{X}_{n+1} - \mathbf{X}(i+1, \mathbf{p}_0) = [\mathbf{A} - \mathbf{B}\mathbf{K}] \cdot [\mathbf{X}_n - \mathbf{X}(i, \mathbf{p}_0)]. \tag{10}$$

For a *controllable* system, standard techniques allow us to find a matrix $\mathbf{K}$ such that $[\mathbf{A} - \mathbf{B}\mathbf{K}]$ has any desired eigenvalues. (The controllability condition and the construction of the matrix $\mathbf{K}$ are described in Barreto and Grebogi, 1995b, and Ogata, 1987.) By selecting eigenvalues of magnitude less than one, $\mathbf{X}_n - \mathbf{X}(i, \mathbf{p}_0)$ approaches 0.

The formulation above assumes that the complete state of the system $\mathbf{X}_n$ is known at any every map iteration. In practice, this is unrealistic. The method can, however, be implemented by measuring a single scalar time series via the embedding technique (Sauer, Casdagli, and Yorke, 1991; Ott, 1993). We refer the interested reader to Ditto, Rauseo, and Spano (1990) and Garfinkel et al., (1992) for further information.

We now emphasize the geometrical aspects of the higher-dimensional algorithm by discussing the two-dimensional case. (We further simplify the discussion by assuming only one control parameter.) A key aspect of higher-dimensional maps is that there exist both stable and unstable directions at each component of an unstable periodic orbit. The stable (unstable) directions are directions along which points approach (leave) the periodic orbit exponentially. The existence of this structure at each point of the trajectory can be seen for the two-dimensional case as follows: Choose a small circle of radius ε around an orbit point $\mathbf{X}(i)$. This circle can be written as $dx^2 + dy^2 = \varepsilon^2$ in the Cartesian coordinate system whose origin is at $\mathbf{X}(i)$. The image of the circle under $\mathbf{F}^{-1}$ can be expressed as $Adx'^2 + Bdx'dy' + Cdy'^2 = 1$, an equation for an ellipse in the Cartesian coordinate system whose origin is at $\mathbf{X}(i-1)$. The coefficients A, B, and C are functions of elements of the inverse Jacobian matrix at $\mathbf{X}(i)$. This deformation from a circle to an ellipse means that the distance along the major axis of the ellipse at $\mathbf{X}(i-1)$ contracts as a result of the map. Similarly the image of a circle at $\mathbf{X}(i-1)$ under $\mathbf{F}$ is typically an ellipse at $\mathbf{X}(i)$, which means that the distance along the inverse image of the major axis of the ellipse at $\mathbf{X}(i)$ expands under $\mathbf{F}$. Thus the major axis of the ellipse at $\mathbf{X}(i-1)$ and the inverse image of the major axis of the ellipse at $\mathbf{X}(i)$ approximate the stable and unstable directions at $\mathbf{X}(i-1)$. We note that typically the stable and unstable directions are not orthogonal to each other, and in rare situations they can be identical (Lai et al., 1993b) (nonhyperbolic dynamical systems).

To calculate these stable and unstable directions, we use an algorithm developed in Lai et al., 1993b. This algorithm can be applied to cases where the period of the orbit is arbitrarily large. To find the stable direction at a point $\mathbf{X}$, we first iterate this point forward N times under the map $\mathbf{F}$ and obtain the trajectory $\mathbf{F}^1(\mathbf{X})$, $\mathbf{F}^2(\mathbf{X}), \ldots, \mathbf{F}^N(\mathbf{X})$. Now imagine that we place a circle of arbitrarily small radius ε at the point $\mathbf{F}^N(\mathbf{X})$. If we iterate this circle backward once, the circle will become an ellipse at the point $\mathbf{F}^{N-1}(\mathbf{X})$, with the major axis along the stable direction of the point $\mathbf{F}^{N-1}(\mathbf{X})$. We continue iterating this ellipse backward, while at the same time rescaling the ellipse's major axis to be order ε. When we iterate the ellipse back to the point $\mathbf{X}$, the ellipse becomes very thin with its major axis along the stable direction at the point $\mathbf{X}$, if N is sufficiently large. For a short period-m orbit, we choose $N = km$ with k an integer. In practice, instead of using a small circle, we take a unit vector at the point $\mathbf{F}^N(\mathbf{X})$, since the Jacobian matrix of the inverse map $\mathbf{F}^{-1}$ rotates a vector in the tangent space of $\mathbf{F}$ toward the stable direction. Hence we iterate a unit vector backward to the point $\mathbf{X}$ by multiplying by the Jacobian matrix of the inverse map at each point on the already existing orbit. We rescale the vector after each multiplication to unit length. For sufficiently large N, the unit vector so obtained at $\mathbf{X}$ is a good approximation to the stable direction at $\mathbf{X}$.

Similarly, to find the unstable direction at point $\mathbf{X}$, we first iterate $\mathbf{X}$ backward under the inverse map N times to obtain a backward orbit $\mathbf{F}^{-j}(\mathbf{X})$ with $j = N, \ldots, 1$. We then choose a unit vector at point $\mathbf{F}^{-N}(\mathbf{X})$ and iterate this unit vector forward to the point $\mathbf{X}$ along the already existing orbit by multiplying by the Jacobian matrix of the map N times. (Recall that the Jacobian matrix of the forward map rotates a vector toward the unstable direction.) We rescale the vector to unit length at each step. The final vector at point $\mathbf{X}$ is a good approximation to the unstable direction at that point if N is sufficiently large.

The method above is efficient. For instance, the error between the calculated and real stable or unstable directions (Lai et al., 1993b) is on the order of 10^{-10} for chaotic trajectories in the Hénon map if $N = 20$.

Let $\mathbf{e}_{s,i}$ and $\mathbf{e}_{u,i}$ be the stable and unstable directions at $\mathbf{X}(i)$, and let $\mathbf{f}_{s,i}$ and $\mathbf{f}_{u,i}$ be the corresponding contravariant vectors that satisfy the conditions $\mathbf{f}_{u,i} \cdot \mathbf{e}_{u,i} = \mathbf{f}_{s,i} \cdot \mathbf{e}_{s,i} = 1$ and $\mathbf{f}_{u,i} \cdot \mathbf{e}_{s,i} = \mathbf{f}_{s,i} \cdot \mathbf{e}_{u,i} = 0$. To stabilize the orbit, we require that the next iteration of a trajectory point, after falling into a small neighborhood about $\mathbf{X}(i)$, fall along the stable direction at $\mathbf{X}(i + 1, p_0)$, namely

$$[\mathbf{X}_{n+1} - \mathbf{X}(i + 1, p_0)] \cdot \mathbf{f}_{u,i+1} = 0. \qquad (11)$$

If we take the dot product of both sides of (8) with $\mathbf{f}_{u,i+1}$ and use (10), we obtain the expression for the parameter perturbations:

$$\Delta p_n = \frac{\{\mathbf{A} \cdot [\mathbf{X}_n - \mathbf{X}(i, \ p_0)]\} \cdot \mathbf{f}_{u,i+1}}{-\mathbf{B} \cdot \mathbf{f}_{u,i+1}}. \qquad (12)$$

The general algorithm for controlling chaos for two-dimensional maps can be summarized as follows:

1. Find the desired unstable periodic orbit to be stabilized.
2. Find a set of stable and unstable directions, $\mathbf{e}_s$ and $\mathbf{e}_u$, at each component of the periodic orbit. The set of corresponding contravariant vectors $\mathbf{f}_s$ and $\mathbf{f}_u$ can be found by solving $\mathbf{e}_s \cdot \mathbf{f}_s = \mathbf{e}_u \cdot \mathbf{f}_u = 1$ and $\mathbf{e}_s \cdot \mathbf{f}_u = \mathbf{e}_u \cdot \mathbf{f}_s = 0$.
3. Randomly choose an initial condition, and evolve the system at the parameter value p_0. When the trajectory enters the ε neighborhood of the target periodic orbit, calculate parameter perturbations at each time step according to (12).

The OGY algorithm described above, when generalized to permit the use of more than one control parameter (Barreto and Grebogi, 1995b), allows greater efficiency in both achieving control and maintaining control in the presence of noise. Furthermore the algorithm is not restricted to the control of unstable periodic orbits. It can be applied to stabilizing chaotic trajectories to synchronize two chaotic systems (Lai and Grebogi, 1993, 1994b) or to convert transient chaos into sustained chaos (Lai and Grebogi, 1994a).

It should be noted that the algorithm discussed above applies to invertible maps. In general, dynamical systems that can be described by a set of first-order autonomous differential equations are invertible, and the inverse system is obtained by letting $t \rightarrow -t$ in the original set of differential equations. Hence the discrete map obtained on the Poincaré surface of section also is invertible. Most dynamical systems encountered in practice fall into this category. Noninvertible dynamical systems possess very distinct properties from invertible dynamical systems (Chossat and Golubitsky, 1988; Chin, Kan, and Grebogi, 1993). For instance, for two-dimensional noninvertible maps a point on a chaotic attractor may not have a unique stable (unstable) direction. A method for determining all these stable and unstable directions is not known. If one or several such directions at the target unstable periodic orbit can be calculated, the OGY method can in principle be applied to noninvertible systems by forcing a chaotic trajectory to fall on one of the stable directions of the periodic orbit.

11.4   TARGETING AND CHAOS CONTROL IN A MECHANICAL SYSTEM

We have seen that the OGY algorithm assumes that the control perturbations are limited to be small, and hence the algorithm is based on a linearization of the dynamics in the immediate vicinity of the unstable orbit that is to be stabilized. As described above, it is possible to rely on the natural ergodicity of chaos to bring the trajectory into this small region before applying the control. However, in the case of even moderately high-dimensional systems, the chaotic transients that occur before the control can be applied may be prohibitively long.

If a reasonable global model of the system of interest is available, it is possible to employ a targeting method to effectively reduce these transient times. (If a global model is not available, similar results may be obtained by using several simultaneous control parameters; see Barreto and Grebogi, 1995b.) Targeting refers to global, nonlocal control of chaos; specifically it is a method by which a chaotic orbit can be rapidly steered to a desired part of the attractor. Several such methods have been proposed (Shinbrot et al., 1990, 1992a, 1992b, 1993; Bollt and Meiss, 1995). We describe here a *tree-targeting* method that is computationally efficient for higher-dimensional systems (Kostelich et al., 1987). When coupled with the OGY algorithm described above, targeting allows for faster and more efficient stabilization.

To demonstrate the advantages of this method, we use the kicked double-rotor system shown in Figure 11.3. This is an idealized physical system of two connected rods subject to periodic impulsive kicks. The time evolution, sampled immediately after each kick, is given by a four-dimensional map (Romeiras et al., 1992; Grebogi et al., 1986, 1987). We take our control

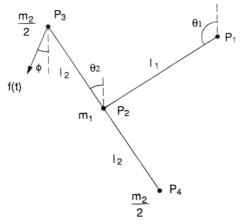

FIGURE 11.3. The kicked double rotor. A massless rod of length l_1 pivots about the stationary point P_1. A second massless rod of length $2l_2$ is mounted on pivot P_2, which in turn is mounted at the end of the first rod. Periodic impulsive kicks $f(t) = \sum_{n=0}^{\infty} \rho_n \delta(t - n)$ are applied at an angle ϕ as shown. The state of the system immediately after the $(n + 1)$th kick is given by a four-dimensional map of the form $\mathbf{X}_{n+1} = \mathbf{M}\mathbf{Y}_n + \mathbf{X}_n$ and $\mathbf{Y}_{n+1} = \mathbf{L}\mathbf{Y}_n + \mathbf{G}(\mathbf{X}_{n+1})$ where $\mathbf{X} = (\theta_1, \theta_2)^T$ are the two angular position coordinates, $\mathbf{Y} = (\dot{\theta}_1, \dot{\theta}_2)^T$ are the corresponding angular velocities, and $\mathbf{G}(\mathbf{X})$ is a nonlinear function. $\mathbf{M}$ and $\mathbf{L}$ are both constant matrices that involve the coefficients of friction at the two pivots and the moments of inertia of the rotor. Gravity is absent. Control parameters at time n are $\rho_n = 9.0 + \delta\rho_n$ and $\phi_n = 0.0 + \delta\phi_n$. We take $l_1 = 1/\sqrt{2}$, and set all other parameters to 1. (For further details, see Romeiras et al., 1992; Grebogi et al., 1986, 1987.)

parameters to be the strength of the kick ρ and the angle ϕ at which the kick is applied. Small perturbations ($|\Delta\rho|/\rho_0 \leqslant 0.1$; $|\Delta\phi| \leqslant 0.5$) are applied around nominal values ($\rho_0 = 9.0$, $\phi_0 = 0$) at which the map exhibits 36 fixed points within a chaotic attractor of Lyapunov dimension 2.8 (see the figure caption for further details). This dimension is to be contrasted with most experimental and numerical studies where the dimension was typically between one and two, and often close to one. Figure 11.4 illustrates the advantages of targeting. We seek to stabilize a sequence of five fixed points in succession. The figure displays the θ_1 component of the state versus iteration number. In Figure 11.4a we rely on ergodicity to bring the orbit close to the desired fixed point before it is stabilized. In Figure 11.4b our tree-targeting method is used. Very large improvements in the switching time is evident; note the great difference in the scales on the two horizontal axes. More precisely, in panel a the switching

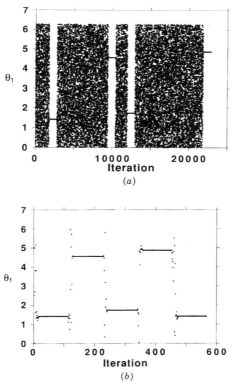

(a)

(b)

FIGURE 11.4. Graphs illustrating switching between five different fixed points. The θ_1 coordinate of the state is plotted against iteration. (a) We rely on ergodicity to bring the orbit close to the desired UPO. The fifth fixed point required 153,485 iterations to be stabilized and is not shown. (b) Tree targeting is employed to significantly reduce transients that precede stabilization.

times t follow an exponential distribution $\langle t \rangle^{-1} \exp(t/\langle t \rangle)$ (this is typical of chaotic systems (Ott, 1993)), and we find that $\langle t \rangle$ ranges from $12{,}000 \pm 80$ iterations to $252{,}000 \pm 3000$ iterations, depending on the fixed point. In sharp contrast, the method outlined below permits the target to be attained in 17–19 iterations using one parameter targeting (ρ), and 13–15 iterations using two parameter targeting (ρ and ϕ). Thus we achieve improvements of 3–4 orders of magnitude in the switching times.

The first step in the tree-targeting procedure is to identify the set of unstable periodic orbits that are to be stabilized. For simplicity, we take these to be fixed points, namely periodic orbits of period one. We denote these by $\mathbf{p}_1, \mathbf{p}_2, \ldots, \mathbf{p}_n$. For each such point we construct *targeting trees*, which function as "road-maps" of the attractor. To stabilize $\mathbf{p}_1$, a chaotic orbit is directed along the corresponding tree into the vicinity of $\mathbf{p}_1$. The OGY method is then applied to stabilize the orbit. To switch to $\mathbf{p}_2$, we can abandon $\mathbf{p}_1$, follow the tree leading to $\mathbf{p}_2$, and subsequently stabilize the orbit there.

Each targeting tree is constructed by first choosing the target, say, the fixed point $\mathbf{p}_1$. The map is then iterated from a random initial condition while keeping in memory a short history of the iterates encountered (e.g., 10 consecutive points) until the orbit lands within a tolerance distance of the target. This point, together with the recorded pre-iterates, comprise the *trunk path* of the tree and are stored in memory. The map is then iterated again, still keeping track of a brief iterate history until the orbit lands near any one of the points already in the tree. When this happens, we add the new path as a *branch*. Continuing in this way, we build a tree with a hierarchy of branches: The trunk path is level 1, level 2 branches are those that are rooted at some point in the trunk path, level 3 branches are rooted at a level 2 branch, and so on. The objective is to build a tree with enough branches such that a typical uncontrolled chaotic orbit lands near a point in the tree after a small number of iterations.

The basic targeting procedure is illustrated in Figure 11.5. Assume that a target point $\mathbf{t} = \mathbf{x}_{10}$ on the attractor has been selected, and that the trunk path consisting of points $\mathbf{x}_9, \mathbf{x}_8, \ldots, \mathbf{x}_0$ has been recorded. Let $\mathbf{y}_0$ be a point near $\mathbf{x}_0$. Without targeting, the orbit $\mathbf{y}_1, \mathbf{y}_2, \ldots$, quickly diverges from the path. We seek a series of small perturbations to available control parameters such that the perturbed orbit $\hat{\mathbf{y}}_1, \hat{\mathbf{y}}_2, \ldots$ lands on the stable manifold of a subsequent point $\mathbf{x}_i$ in the path. If this can be accomplished for a small value of i, then the orbit will quickly approach the path, and $\hat{\mathbf{y}}_{10}$ will be very close to the target $\mathbf{x}_{10}$.

For specificity, we refer to the case of the kicked double rotor (Figure 11.3, i.e., a four-dimensional map $\mathbf{F}$ with two positive and two negative Lyapunov exponents). Let S_2 represent the (typically two-dimensional) stable manifold associated with the point $\mathbf{x}_2$. For simplicity we assume that only one parameter ρ is available for control. Recall that two two-planes generically intersect at a single point in $\mathcal{R}^4$. Hence the vectors $\mathbf{g}_1 = \partial \mathbf{F}(\mathbf{F}(\mathbf{y}_0, \rho_1), \rho_2)/\partial \rho_1$ and $\mathbf{g}_2 = \partial \mathbf{F}(\mathbf{F}(\mathbf{y}_0, \rho_1), \rho_2)/\partial \rho_2$ typically span a two-plane through $\hat{\mathbf{y}}_2$ that intersects the two-dimensional stable manifold S_2 at a unique point $\hat{\mathbf{y}}_2$. Therefore we look

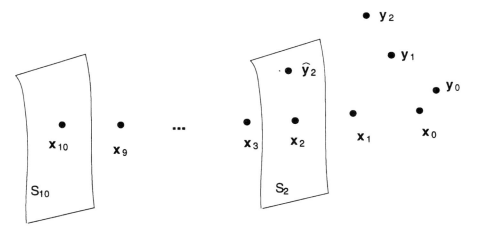

FIGURE 11.5. Schematic of a targeting procedure. Two successive perturbations of the kick are applied at y_0 to steer it onto the stable manifold associated with the point x_2.

for two successive parameter perturbations ρ_1 and ρ_2 such that $\hat{y}_2$ lies on S_2.

The intersection point $\hat{y}_2$ may not be sufficiently close to x_2 to justify using the linear approximation for estimating S_2. This is generally the case in our kicked double-rotor example. Alternatively, we can estimate the intersection point on S_2 by calculating the inverse images of points near subsequent points further down the path. Let s_1 and s_2 denote vectors that span a plane at x_8. A point $z = x_8 + \sigma_1 s_1 + \sigma_2 s_2$ is chosen with σ_1 and σ_2 small (typically of order 10^{-6}). The inverse images $F^{-1}(z)$, $F^{-2}(z),\dots$ rapidly approach the stable manifolds $S_7, S_6,\dots$ because, under the inverse map, S_8 is an expanding set and components perpendicular to S_8 contract.

Thus in the case of one parameter control, we calculate two parameter perturbations ρ_1 and ρ_2, together with values for σ_1 and σ_2, such that

$$F^{-k}(x_{k+2} + \sigma_1 s_1 + \sigma_2 s_2) = F(F(y_0, \rho_1), \rho_2) = \hat{y}_2. \qquad (13)$$

In our example, we use $k = 6$. Equation (13) can be solved numerically using Newton's method. Once the prescribed kicks ρ_1 are ρ_2 are applied at y_0 and $\hat{y}_1$, the orbit lands on the stable manifold of x_2 (at $\hat{y}_2$), and subsequent iterations of the map approach the path exponentially.

In practice, values of k which yield numerically accurate results can be determined by performing numerical trials on the particular map being considered. To correct for these and other nonideal effects such as noise, state measurement error, and an imperfect determination of the system parameters, the method is reapplied at every iteration.

If two parameters ρ and ϕ are available for control, then only one perturbation step is necessary: Typically there is a two-plane through y_1

spanned by $\mathbf{g}_\rho = \partial \mathbf{F}(\mathbf{y}_0, \rho, \phi)/\partial \rho$ and $\mathbf{g}_\phi = \partial \mathbf{F}(\mathbf{y}_0, \rho, \phi)/\partial \phi$ that intersects the stable manifold S_1 of $\mathbf{x}_1$. The procedure outlined above can be similarly extended to different dimensions and different numbers of positive Lyapunov exponents.

Assume now that a three-level targeting tree has been constructed. The map can be iterated until the trajectory lands at a point $\mathbf{y}$ near a point $\mathbf{x}$ in the tree. Suppose that $\mathbf{x}$ is in a level 3 branch. The base of this branch is chosen as an interim target, and the orbit is directed there by the method described above. Next we set the interim target to be the root of the adjoining level 2 branch. The orbit is steered to this new target, and the process is repeated until the final target is attained. The orbit is then stabilized at the fixed point by applying the OGY control procedure.

In the procedure described above, an initial condition $\mathbf{z}$ is iterated until the uncontrolled orbit encounters the tree. Another possibility is to generate a cloud of points by calculating the image of $\mathbf{z}$ under a small random parameter perturbation (applied to ρ) and repeating this many times. Thus all points in the cloud are within one iteration of $\mathbf{z}$. This entire cloud can then be iterated forward a certain number of times, and each time a point in the cloud encounters the tree, its position is recorded. In this way many different paths from the initial condition $\mathbf{z}$ onto the tree can be found, and from these we can select the path that ultimately reaches the target in the fewest number of iterations.

11.5 CONCLUSIONS

In summary, we have presented an algorithm for converting chaos into periodic motion by using only small parameter perturbations. This method takes advantage of chaos and thus avoids the need to make large-scale changes to the system of interest. We have also described a global control method suitable for higher-dimensional systems that efficiently reduces the length of chaotic transients that precede stabilization. These methods promise to have a profound impact on material processing and manufacturing.

ACKNOWLEDGMENTS

This work was supported by DOE (Office of Scientific Computing, Office of Energy Research). In addition E. B. was supported by the National Physical Science Consortium under the sponsorship of Argonne National Laboratory.

REFERENCES

Auerbach, D., Grebogi, C., Ott, E., et al. 1992. Controlling chaos in high-dimensional systems. *Phys. Rev. Lett.* **69**:3479–3482.

Barreto, E., Kostelich, E. J., Grebogi, C., et al. 1995a. Efficient switching between controlled unstable periodic orbits in higher dimensional chaotic systems. *Phys. Rev. E* **51**:4169–4172.

Barreto, E., and Grebogi, C. 1995b. Multiparameter control of chaos. *Phys. Rev. E* **52**:3553–3557.

Bielawski, S., Bouazaoui, M., Derozier, D., et al. 1993. Experimental characterization of unstable periodic orbits by controlling chaos. *Phys. Rev. A* **47**:3276–3279.

Bielawski, S., Derozier, D., and Glorieux, P. 1994. Controlling unstable periodic orbits by a delayed continuous feedback. *Phys. Rev. E* **49**:R971–R794.

Bollt, E., and Meiss, J. D. 1995. Controlling chaotic transport through recurrence. *Physica D* **81**:280–294.

Chin, W., Kan, I., and Grebogi, C. 1993. Evolution of attractor boundaries in two-dimensional noninvertible maps. *Random Comp. Dynamics* **1**:349–370.

Chossat, P., and Golubitsky, M. 1988. Symmetry-increasing bifurcation of chaotic attractors. *Physica D* **32**:423–436.

Daw, C. S., Finney, C. E. A., Vasudevan, M., et al. 1995. Self-organization and chaos in a fluidized bed. *Phys. Rev. Lett.* **75**:2308–2311.

Ditto, W. L., Rauseo, S. N., and Spano, M. L. 1990. Experiment control of chaos. *Phys. Rev. Lett.* **65**:3211–3214.

Dressler, U., and Nitsche, G. 1992. Controlling chaos using time delay coordinates. *Phys. Rev. Lett.* **68**:1–4.

Feigenbaum, M. J. 1978. Quantitative university for a class of nonlinear transformations. *J. Stat. Phys.* **19**:25–52.

Garfinkel, A., Spano, M. L., Ditto, W. L., et al. 1992. Controlling cardiac chaos. *Science* **257**:1230–1235.

Gauthier, D. J., Sukow, D. W., Concannon, H. M., et al. 1994. Stabilizing unstable periodic orbits in a fast diode resonator using continuous time-delay autosynchronization. *Phys. Rev. E* **50**:2343–2346.

Gills, Z., Iwata, C., Roy, R., et al. 1992. Tracking unstable steady states: Extending the stability regime of a multimode laser system. *Phys. Rev. Lett.* **69**:3169–3172.

Grebogi, C., Kostelich, E., Ott, E., et al. 1986. Multi-dimensional intertwined basin boundaries and the kicked double rotor. *Phys. Lett. A* **118**:448–452; Erratum (1987) **120A**:497.

Grebogi, C., Kostelich, E. Ott, E., et al. 1987. Multi-dimensional intertwined basin boundaries: Basin structure of the kicked double rotor. *Physica D* **25**:347–360.

Hayes, S., Grebogi, C., and Ott, E. 1993. Communicating with chaos. *Phys. Rev. Lett.* **70**:3031–3034.

Hayes, S., Grebogi, C., Ott, E., et al. 1994. Experimental control of chaos for communication. *Phys. Rev. Lett.* **73**:1781–1784.

Hübinger, B., Doerner, R., Martienssen, W., et al. 1994. Controlling chaos experimentally in systems exhibiting large effective Lyapunov exponents. *Phys. Rev. E* **50**:932–948.

Hunt, E. R. 1991. Stabilizing high-period orbits in a chaotic system: The diode resonator. *Phys. Rev. Lett.* **67**:1953–1955.

Johnson, G. A., and Hunt, E. R. 1993. Controlling in a simple autonomous system: Chua's circuit. *Int. J. Bifurcation Chaos* **3**:789–792.

Kostelich, E., Grebogi, C., Ott, E., et al. 1993. Higher-dimensional targeting. *Phys. Rev. E* **47**:305–310.

Lai, Y. C., and Grebogi, C. 1993. Synchronization of chaotic trajectories using control. *Phys. Rev. E* **47**:2357–2360.

Lai, Y. C., and Grebogi, C. 1994a. Converting transient chaos into sustained chaos by feedback control. *Phys. Rev. E* **49**:1094–1098.

Lai, Y. C., and Grebogi, C. 1994b. Synchronization of spatiotemporal chaotic systems by feedback control. *Phys. Rev. E* **50**:1894–1899.

Lai, Y. C., Ding, M., and Grebogi, C. 1993a. Controlling Hamiltonian chaos. *Phys. Rev. E* **47**:86–92.

Lai, Y. C., Grebogi, C., Yorke, J. A., et al. 1993b. How often are chaotic saddles nonhyperbolic. *Nonlinearity* **6**:779–797.

Lai, Y. C., Tél, T., and Grebogi, C. 1993c. Stabilizing chaotic-scattering trajectories using control. *Phys. Rev. E* **48**:709–717.

Meucci, R., Gadomski, W., Ciofini, M., et al. 1994. Experimental control of chaos by means of weak parametric perturbations. *Phys. Rev. E* **49**:R2528–R2531.

Moon, F. C., Johnson, M. A., and Holmes, W. T. 1996. Controlling chaos in a two-well oscillator. *Int. J. Bifurcation Chaos* **6**:337–347.

Ogata, K. 1987. *Discrete-Time Control Systems*. Prentice Hall, Englewood Cliffs, NJ.

Ott, E. 1993. *Chaos in Dynamical Systems*. Cambridge University Press, Cambridge.

Ott, E., Grebogi, C., and Yorke, J. A. 1990. Controlling chaos, *Phys. Rev. Lett.* **64**:1196–1199.

Petrov, V., Crowley, M. J., and Showalter, K. 1994. Tracking unstable periodic orbits in the Belousoc-Zhabotinsky reaction. *Phys. Rev. Lett.* **72**:2955–2958.

Petrov, V., Gaspar, V., Masare, J., et al. 1993. Controlling chaos in the Belousov-Zhabotinsky reaction. *Nature* **361**:240–243.

Reyl, C., Flepp, L., Badii, R., et al. 1993. Control of NMR-laser chaos in high-dimensional embedding space. *Phys. Rev. E* **47**:267–272.

Rhode, M. A., Rollins, R. W., Markworth, A. J., et al. 1995. Controlling chaos in a model of thermal pulse combustion. *J. Appl. Phys.* **78**:2224–2232.

Romeiras, F. J., Grebogi, C., Ott, E., et al. 1992. Controlling chaotic dynamical systems. *Physica D* **58**:165–192.

Roy, R., Murphy, T. W., Maier, T. D., et al. 1992. Dynamical control of a chaotic laser: Experimental stabilization of a globally coupled system. *Phys. Rev. Lett.* **68**:1259–1262.

Sauer, T., Casdagli, M., and Yorke, J. A. 1991. Embedology. *J. Stat. Phys.* **65**:579–616.

Schiff, S. J., Jerger, K., Duong, D. H., et al. 1994. Controlling chaos in the brain. *Nature* **370**:615–620.

Shinbrot, T., Ott, E., Grebogi, C., et al. 1990. Using chaos to direct trajectories to targets. *Phys. Rev. Lett.* **65**:3215–3218.

Shinbrot, T., Ott, E., Grebogi, C., et al. 1992a. Using chaos to direct orbits to targets in systems describable by a one-dimensional map. *Phys. Lett. A* **45**:4165–4168.

Shinbrot, T., Ditto, W., Grebogi, C., et al. 1992b. Using the sensitive dependence of chaos (the "butterfly effect") to direct trajectories in an experimental chaotic system. *Phys. Rev. Lett.* **68**:2863–2866.

Shinbrot, T., Grebogi, C., Ott, E., et al. 1993. Using small perturbations to control chaos. *Nature* **363**:411–417.

So, P., and Ott, E. 1995. Controlling chaos using time delay coordinates via stabilization of periodic orbits. *Phys. Rev. E* **51**:2955–2962.

Starrett, J., and Tagg, R. 1994. Control of a chaotic parametrically driven pendulum. *Phys. Rev. Lett.* **74**:1974–1977.

Tél, T. 1991. Controlling transient chaos. *J. Phys. A.* **24**:1359–1368.

12

EXPERIMENTAL
CONTROL OF
HIGH-DIMENSIONAL
CHAOS

DAVID J. CHRISTINI, JAMES J. COLLINS,
and PAUL S. LINSAY

The widespread existence of chaotic dynamics in physical systems has fostered great interest in the development of practical chaos control techniques. The original feedback chaos control technique developed by Ott, Grebogi, and Yorke (OGY) (Ott et al., 1990) is based on the fact that there are an infinite number of unstable periodic orbits (UPOs) embedded within a chaotic attractor. The OGY approach exploits the sensitivity of chaos to initial conditions by making small time-dependent perturbations to an accessible systemwide parameter such that the system's state point is attracted toward the stable direction of a targeted UPO. The OGY technique is practical from an experimental standpoint because it requires no analytical model of the system—all necessary dynamics are estimated from time series measurements made on the system.

The OGY approach and similar model-independent feedback control techniques have been successfully used to control a wide range of experimental systems (Ditto et al., 1990; Hunt, 1991; Peng et al., 1991; Roy et al., 1992; Petrov et al., 1993). However, because the OGY technique is limited to the

Dynamics and Chaos in Manufacturing Processes, Edited by Francis C. Moon.
ISBN 0-471-15293-5 © 1998 John Wiley & Sons, Inc.

control of low-dimensional systems, it is not applicable to the majority of real-world (i.e., high-dimensional) systems. Several high-dimensional control algorithms (Auerbach et al., 1992; Romeiras et al., 1992; So and Ott, 1995), including one recently used to control a magnetoelastic ribbon in a state of high-dimensional chaos (In et al., 1996; Ding et al., 1996), have been developed to overcome this limitation. However, because these techniques, like the OGY approach, apply control perturbations only once per period, amplification of noise and measurement errors by highly unstable systems may lead to control failure (Hübinger et al., 1993). Recently a quasi-continuous chaos control technique known as the *local control method* (Hübinger et al., 1993, 1994) has been developed to reduce the likelihood of control failure by applying several control perturbations per period. This model-independent technique has been successfully used to control two low-dimensional chaotic systems: a driven single pendulum (Hübinger et al., 1994; Jan de Korte et al., 1995) and a driven bronze ribbon (Hübinger et al., 1994). In this study we extend the local control method to stabilize a high-dimensional chaotic system: the driven double pendulum.

During each drive period the local control method attempts to stabilize a targeted UPO by applying N control perturbations δp to an accessible parameter p, such that $p = \bar{p} + \delta p$, where $\bar{p}$ is the initial value of p. To determine the perturbations required to stabilize the targeted UPO, this method introduces N successive Poincaré sections $\Sigma^n, n = 0, \dots (N - 1)$, where Σ^n is intersected by $\mathbf{z}^n$, the nth system state vector. The local control method developed in Hübinger et al. (1993, 1994) utilizes a state vector that is entirely comprised of measured variables. Here we extend the method by using time delay coordinates,[1] where $\mathbf{z}^n$ is comprised of both current and former values of measured variables. With time-delay coordinates, $\mathbf{z}^{n+1}$ is a function of $\mathbf{z}^n$, and all values of p during the delay lag m (i.e., $p^n, p^{n-1}, \dots, p^{n-m+1}$) (Dressler and Nitsche, 1992). Thus the mapping $\mathbf{P}^{(n,n+1)}$ from Σ^n to Σ^{n+1} is

$$\mathbf{z}^{n+1} = \mathbf{P}^{(n,n+1)}(\mathbf{z}^n, p^n, p^{n-1}, \dots, p^{n-m+1}). \tag{1}$$

Letting $\delta\mathbf{z}^n = \mathbf{z}^n - \mathbf{z}_F^n$, where $\mathbf{z}_F^n \in \Sigma^n$ is the intersection of the UPO with Σ^n, the linear approximation of $\mathbf{P}^{(n,n+1)}$ around $\mathbf{z}_F^n$ and $\bar{p}$ gives

$$\delta\mathbf{z}^{n+1} = \mathbf{A}^n\delta\mathbf{z}^n + \sum_{j=0}^{m-1} \mathbf{w}_j^n\delta p^{n-j}, \tag{2}$$

where

$$\mathbf{A}^n = D_{\mathbf{z}^n}\mathbf{P}^{(n,n+1)}(\mathbf{z}_F^n, \bar{p}), \tag{3}$$

$$\mathbf{w}_j^n = \frac{\delta\mathbf{P}^{(n,n+1)}(\mathbf{z}_F^n, \bar{p})}{\delta p^{n-j}}. \tag{4}$$

---

[1] In a different manner, Jan de Korte et al. (1995) also extends the local control method by using time-delay coordinates.

The Jacobian matrix $\mathbf{A}^n$ represents the linearization of $\mathbf{P}^{(n,n+1)}$ around $\mathbf{z}_F^n$, while $\mathbf{w}_j^n$ are vectors that measure the sensitivity of $\mathbf{P}^{(n,n+1)}$ to current $(j=0)$ and former $(j>0)$ parameter perturbations.

Computation of the local control parameter perturbation δp utilizes the fact that $\mathbf{A}^n$ deforms a hypersphere surrounding $\mathbf{z}_F^n$ in Σ^n into a hyperellipsoid surrounding $\mathbf{z}_F^{n+1}$ in Σ^{n+1}. Singular value decomposition of $\mathbf{A}^n$ ($\mathbf{A}^n = \mathbf{U}^n \mathbf{W}^n \mathbf{V}^{nT}$, where superscript T denotes transpose) is used to obtain the Σ^n hypersphere vector $\mathbf{v}_u^n$ (the direction of maximal stretching), which is mapped onto the largest axis of the Σ^{n+1} hyperellipsoid. Thus the vector $\mathbf{v}_u^n$ is the column vector of $\mathbf{V}^n$ corresponding to the largest singular value μ_u^n of $\mathbf{W}^n$. Once $\mathbf{z}^n$ enters into the hypersphere neighborhood surrounding $\mathbf{z}_F^n$, the local control method attempts to constrain the system within the target UPO by selecting δp^n such that the projection of $\delta\mathbf{z}$ onto $\mathbf{v}_u$ decreases by a factor of $(1-\rho)$ during each control step, namely

$$\mathbf{v}_u^{n+1\,T}\delta\mathbf{z}^{n+1} = (1-\rho)\mathbf{v}_u^{n\,T}\delta\mathbf{z}^n. \tag{5}$$

Thus the local control formula is obtained by inserting Equation (5) into (2):

$$\delta p^n = \frac{(1-\rho)\mathbf{v}_u^{n\,T}\delta\mathbf{z}^n - \mathbf{v}_u^{n+1\,T}(\mathbf{A}^n\delta\mathbf{z}^n + \Sigma_{j=0}^{m-1}\mathbf{w}_j^n\delta p^{n-j})}{\mathbf{v}_u^{n+1\,T}\mathbf{w}_0^n}. \tag{6}$$

As in Hübinger et al. (1994), we limit the applied perturbation to $|\delta p^n| \leqslant \delta p_{\max}$,[2] namely

$$\delta p^n = \begin{cases} \delta p^n, & \text{for } |\delta p^n| \leqslant \delta p_{\max}, \\ \text{sign}(\delta p^n)\delta p_{\max}, & \text{for } |\delta p^n| > \delta p_{\max}. \end{cases} \tag{7}$$

The setup for the driven double-pendulum control experiment is shown in Figure 12.1. The electric drive motor is powered by a sinusoidal voltage $V(t) = A\sin(2\pi f t) + p$. For this experiment $A = 1.3\,\text{V}$, $f = 1.2\,\text{Hz}$, and p is a DC torque that is used as the accessible control parameter ($\bar{p} = 0.0\,\text{V}$). The angular velocity $\dot{\theta}_i$ of the inner pendulum is measured by an electronic circuit connected to the voltage output of a generator, whose axle is linked to the axle of the drive motor. The electronic circuit then integrates $\dot{\theta}_i$ to provide the inner pendulum angle θ_i. Whenever the pendulum swings through $\theta_i = 0$, a vertical line on the back of the inner pendulum arm is detected by a bar-code reader. The bar-code reader triggers an integration reset ($\theta_i = 0$) to constrain $-2\pi < \theta_i < 2\pi$. The $\dot{\theta}_i$, θ_i, and V signals are scanned via A/D conversion into a PowerMacintosh 7100/80 computer at a sampling rate of Nf ($N = 20$), the

---

[2]$\delta p_{\max}$ dictates how close $\mathbf{z}^n$ must come to $\mathbf{z}_F^n$ (i.e., the size of the hypersphere neighborhood) before control can be initiated.

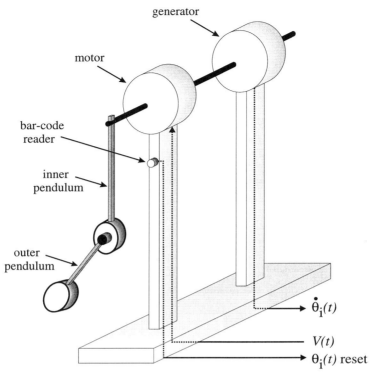

FIGURE 12.1. Schematic of the driven double-pendulum setup. The inner and outer pendulums are aluminum bars weighted on one end (distal end). Each pendulum has full rotational freedom about an axis passing through its proximal end. The outer pendulum's rotational axis passes through the center of the inner pendulum weight and is in the same direction as an electric motor axle that serves as the inner pendulum rotational axis. The electric motor generates an external torque that is dependent on a sinusoidal drive voltage $V(t) = A\sin(2\pi ft) + p$. The angular velocity $\dot{\theta}_i$ of the inner pendulum is measured by a generator whose axle is linked to the axle of the motor. The $\dot{\theta}_i$ signal is integrated by an electronic circuit (not shown) to provide the inner pendulum angle θ_i.

same rate at which control perturbations δp are returned via D/A conversion from the computer to the drive motor.

To locate the target UPO on which control was to be attempted, $\dot{\theta}_i$ and θ_i were recorded from the driven double pendulum for 7500 drive cycles with $\delta p = 0.0$ V. The driven double pendulum has 5 degrees of freedom (the drive phase, $\dot{\theta}_i$, θ_i, $\dot{\theta}_o$, and θ_o, where $\dot{\theta}_o$ and θ_o represent the angular velocity and angle of the outer pendulum, respectively). In this experiment, $\dot{\theta}_o$ and θ_o were not available via measurement. Thus we used a time-delay coordinate embedding comprised of two $\{\dot{\theta}_i, \theta_i\}$ pairs, to reconstruct the pendulum dynamics: $\mathbf{z}^n = (\dot{\theta}_i^n, \theta_i^n, \dot{\theta}_i^{n-m}, \theta_i^{n-m})^T$, where $m = 5$.[3] The method of false nearest-neighbors

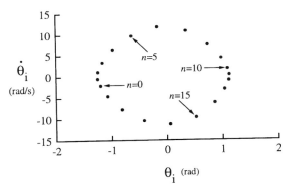

FIGURE 12.2. The libration unstable periodic orbit (UPO), shown in the θ_i–$\dot\theta_i$ plane, for the driven double pendulum of Figure 12.1.

(Kennel et al., 1992) indicated that using three $\{\dot\theta_i, \theta_i\}$ pairs for the time-delay embedding produced the optimal attractor reconstruction (false nearest-neighbor percentage = 8.0%). However, this was only a 2.7% improvement over using two $\{\dot\theta_i, \theta_i\}$ pairs for the time-delay embedding (false nearest-neighbor percentage = 10.7%). Consequently we used two $\{\dot\theta_i, \theta_i\}$ pairs for the time-delay embedding in order to reduce the complexity of the control intervention computations (Equation (6)). The entire data record was searched for period-1 orbits, namely segments of the time series that satisfied $|(\dot\theta_i^n - \dot\theta_i^{n+N})/\dot\theta_i^n| < 0.025$ and $|(\theta_i^n - \theta_i^{n+N})/\theta_i^n| < 0.025$. Since more than one unique UPO can exist for a given period, each period-1 orbit was classified as either (1) a unique orbit or (2) a recurrence of a previously identified orbit. An orbit was considered recurrent if it had the same drive phase as a previously identified orbit and if each of its $\dot\theta_i^n$ and θ_i^n were within 5% of those of the previously identified orbit. Each orbit with at least 10 recurrences was considered to be a valid UPO. The components z_F^n of each valid UPO were computed as the averages of all recurrences of $\dot\theta_i^n$, θ_i^n, $\dot\theta_i^{n-m}$, and θ_i^{n-m}. Figure 12.2 shows the double pendulum's libration UPO, which was selected for control in this study because it had more recurrences (215 recurrences) than any other UPO.

The Jacobian matrices and sensitivity vectors of Equation (6) were estimated from a second recording during which a single δp perturbation was applied each period.[4] The duration of each perturbation was $1/(Nf)$ (i.e., 0.04 s, which was the same duration used during control), and each amplitude was

---

[3]This value was selected because the first minimum in the mutual information (Fraser and Swinney, 1986) for $\dot\theta_i$ occurred at five samples.

[4]In previous local control (Hübinger et al., 1994; Jan de Korte et al., 1995) applications, a different approach was employed: One recording (with no perturbations) was used for Jacobian estimation, and a separate recording (with perturbations) was used to estimate the sensitivity vectors.

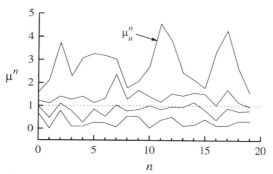

FIGURE 12.3. Singular values μ^n of $\mathbf{A}^n$ for the UPO shown in Figure 12.2. Values of $\mu > 1.0$ indicate an unstable (expanding) direction, while values of $\mu < 1.0$ indicate a stable (contracting) direction. The largest singular value μ_u^n corresponds to the direction of maximum stretching.

randomly selected as $-0.3, -0.15, 0.0, +0.15,$ or $+0.3$ V. During the recording the application of the perturbations was timed such that each $n \in N$ received 3500 perturbations. After the recording the Jacobian matrices and sensitivity vectors were estimated from the nearest neighbors of each $\mathbf{z}_F^n$. Because the sensitivity vectors are only affected by perturbations that occur during the delay lag m, the nearest-neighbors search was limited to vectors that followed an applied perturbation by fewer than m lags. For each $n \in N$ the 40 nearest neighbors of $\mathbf{z}_F^n$, their corresponding $\mathbf{z}_F^{n+1}$, and the corresponding perturbations δp^{n-j} (where j equals the number of scans between the perturbation and the nearest neighbor) were simultaneously fit to Equation (2) to estimate $\mathbf{A}^n$ and $\mathbf{w}_j^n$ ($j = 0, 1, \ldots, m - 1$).

The Lyapunov numbers ($\lambda_1 = 138.0$, $\lambda_2 = -2.8$, $\lambda_3 = 0.3$, and $\lambda_4 = 1.0 \times 10^{-4}$)[5] for the target UPO shown in Figure 12.2 indicate that the orbit has two unstable directions.[6] Figure 12.3 shows the singular values μ^n for each $\mathbf{A}^n$ of the target UPO. The variation of the singular values around the orbit indicates that orbit stability is dependent on n. For each $n \in N$, there are at least two unstable directions ($\mu > 1.0$). It should also be noted that each of the $\mathbf{A}^n$ Jacobian matrices estimated from Equation (2) was characterized by at least two complex eigenvalues. Thus a local control approach that employs eigenvalues, rather than singular value decomposition, would be inappropriate for this system and other systems with complex eigenvalues for $\mathbf{A}^n$.

---

[5]The Lyapunov numbers were computed by a method utilizing QR decomposition, as described in Eckmann and Ruelle (1985).

[6]Although the control law of Equation (6) considers only the largest unstable direction, the magnitude of the second unstable direction (characterized by λ_2) was not large enough to cause control failure.

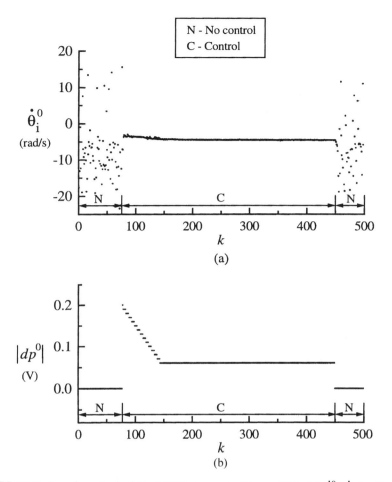

FIGURE 12.4. Local control of the UPO shown in Figure 12.2. (a) $\dot{\theta}_i^0$ ($\dot{\theta}_i$ for the first Poincaré section Σ^0) versus cycle number k. (b) Absolute value of the corresponding control perturbations δp^0 computed by Equations (6) and (7) with $\rho = 0.15$. The duration of each perturbation was 0.04 s. Control was inactive ($\delta p = 0$) until $\mathbf{z}^n$ entered into the hypersphere neighborhood surrounding $\mathbf{z}_F^n$ at $k = 77$. Control was then activated with $\delta p_{max} = 0.2$ V. Subsequently δp_{max} was uniformly decreased to $\delta p_{max} = 0.06$ V. At $k = 450$ control was turned off, and the double pendulum resumed its chaotic motion. The respective control stages are annotated in (a) and (b).

Figure 12.4 shows local control of the inner pendulum angular velocity, along with the corresponding parameter perturbations, for the target UPO in Figure 12.2. Control initiation occurred when $\mathbf{z}^n$ entered into the hypersphere surrounding $\mathbf{z}_F^n$. The initial size of the hypersphere (corresponding to a maximum allowable parameter perturbation of $\delta p_{max} = 0.2$ V) was selected to allow a timely entry into the $\mathbf{z}_F^n$ neighborhood. Once control was obtained, δp_{max} was uniformly decreased to a value of approximately 5% of the drive amplitude A. Control could be maintained indefinitely, even with these small perturbations. The stabilized orbit never exactly matched the target UPO. (This was consistent with Hübinger et al., 1994.) Thus the control perturbations computed by Equation (6) consistently exceeded δp_{max} and were capped by Equation (7). Once control was turned off, the double pendulum quickly resumed its chaotic motion (Figure 12.4).

To demonstrate the robustness of control, "measurement" noise was added to each component of $\mathbf{z}^n$, namely

$$\mathbf{z}^n = (\dot{\theta}_i^n + \varepsilon\xi_1^n, \; \theta_i^n + \varepsilon\xi_2^n, \; \dot{\theta}_i^{n-m} + \varepsilon\xi_3^n, \; \theta_i^{n-m} + \varepsilon\xi_4^n)^T,$$

where $\xi_1^n, \xi_2^n, \xi_3^n$, and ξ_4^n are independent random variables uniformly distributed in $[-1, 1]$ and ε is a constant. At $\delta p_{max} = 0.06 \, \text{V}$, control could be maintained indefinitely for $\varepsilon = 0.05A$. Interestingly control could also be maintained (in the absence of additive noise) even if former perturbations were excluded from Equation (6), namely $w_j^n = 0$ for $j > 0$, thus further indicating the robustness of the local control method.

In this study we have demonstrated that a delay coordinate extension of the local control method can be used to control an experimental high-dimensional chaotic system. These developments may further open up real-world applications of chaos control. For example, the approach used in the present study may be particularly appropriate for mechanical and biological (Garfinkel et al., 1992; Schiff et al., 1994; Christini and Collins, 1995, 1996) systems, which are often high-dimensional.

ACKNOWLEDGMENTS

This work was supported by the National Science Foundation (JJC, DJC) and the Office of Naval Research (PSL).

REFERENCES

Auerbach, D., Grebogi, C., Ott, E., and Yorke, J. A. 1992. Controlling chaos in high dimensional systems. *Phys. Rev. Lett.* **69**:3479–3482.

Christini, D. J., and Collins, J. J. 1995. Controlling nonchaotic neuronal noise using chaos control techniques. *Phys. Rev. Lett.* **75**:2782–2785.

Christini, D. J., and Collins, J. J. 1996. Using chaos control and tracking to suppress a pathological nonchaotic rhythm in a cardiac model. *Phys. Rev. E* **53**:R49–R52.

Ding, M., Yang, W., In, V., Ditto, W. L., Spano, M. L., and Gluckman, B. J. 1996. Controlling chaos in high dimensions: Theory and experiment. *Phys. Rev. E* **53**:4334–4344.

Ditto, W. L., Rauseo, S. N., and Spano, M. L. 1990. Experimental control of chaos. *Phys. Rev. Lett.* **65**:3211–3214.

Dressler, U., and Nitsche, G. 1992. Controlling chaos using time delay coordinates. *Phys. Rev. Lett.* **68**:1–4.

Eckmann, J. P., and Ruelle, D. 1985. Ergodic theory of chaos and strange attractors. *Rev. Mod. Phys.* **57**:617–656.

Fraser, A. M., and Swinney, H. L. 1986. Independent coordinates for strange attractors from mutual information. *Phys. Rev. A* **33**:1134–1140.

Garfinkel, A., Spano, M. L., Ditto, W. L., and Weiss, J. N. 1992. Controlling cardiac chaos. *Science* **257**:1230–1235.

Hübinger, B., Doerner, R., and Martienssen, W. 1993. Local control of chaotic motion. *Z. Phys. B* **90**:103–106.

Hübinger, B., Doerner, R., Martienssen, W., Herdering, W., Pitka, R., and Dressler, U. 1994. Controlling chaos experimentally in systems exhibiting large effective Lyapunov exponents. *Phys. Rev. E* **50**:932–948.

Hunt, E. R. 1991. Stabilizing high-period orbits in a chaotic system: The diode resonator. *Phys. Rev. Lett.* **67**:1953–1955.

In, V., Ditto, W. L., Ding, M., Yang, W., Spano, M. L., and Gluckman, B. J. 1997. Experimental control of high dimensional chaos. *Preprint.*

Jan de Korte, R., Schouten, J. C., and van den Bleek, C. M. 1995. Experimental control of a chaotic pendulum with unknown dynamics using delay coordinates. *Phys. Rev. E* **52**:3358–3365.

Kennel, M. B., Brown, R., and Abarbanel, H. D. I. 1992. Determining embedding dimension for phase-space reconstruction using a geometrical construction. *Phys. Rev. A* **45**:3403–3411.

Ott, E., Grebogi, C., and Yorke, J. A. 1990. Controlling chaos. *Phys. Rev. Lett.* **64**:1196–1199.

Peng, B., Petrov, V., and Showalter, K. 1991. Controlling chemical chaos. *J. Phys. Chem.* **95**:4957–4959.

Petrov, V., Gáspár, V., Masere, J., and Showalter, K. 1993. Controlling chaos in the Belousov-Zhabotinsky reaction. *Nature* **361**:240–243.

Romeiras, F. J., Grebogi, C., Ott, E., and Dayawansa, W. P. 1992. Controlling chaotic dynamical systems. *Physica D* **58**:165–192.

Roy, R., Murphy, Jr., T. W., Maier, T. D., Gills, Z., and Hunt, E. R. 1992. Dynamical control of a chaotic laser: Experimental stabilization of a globally coupled system. *Phys. Rev. Lett.* **68**:1259–1262.

Schiff, S. J., Jerger, K., Duong, D. H., Chang, T., Spano, M. L., and Ditto, W. L. 1994. Controlling chaos in the brain. *Nature* **370**:615–620.

So, P., and Ott, E. 1995. Controlling chaos using time delay coordinates via stabilization of periodic orbits. *Phys. Rev. E* **51**:2955–2962.

INDEX

313